W9-CZM-289

ERGONOMICS

A PRACTICAL GUIDE

2nd Edition

National Safety Council

Project editor: Patricia M. Laing
Technical advisers: Barbra Jean Dembski
              Jill Niland
Cover design: Bob Sunyog
Composition and interior design: Publishing Solutions

Library of Congress Cataloging in Publication Data
National Safety Council
Ergonomics: A Practical Guide
International Standard Book Number: 0-87912-168-8
Library of Congress Catalog Card Number: 92-085026
1M1092   Product Number: 12218-0000

Contents

Preface

ERGONOMICS IS A BODY OF KNOWLEDGE about human abilities, human limitations, and other human characteristics that are relevant to design. Ergonomics, as used in the workplace, seeks to apply this body of knowledge to the design of tools, machines, systems, tasks, jobs, and environments for safe, comfortable, and effective human use. In other words, ergonomics seeks to fit the job to the worker to prevent the development of occupational injuries or illnesses and to reduce the potential for fatigue, error, or unsafe practices.

Ergonomics: A Practical Guide is a manual to help safety and health professionals identify and correct ergonomic problems in their workplace. It discusses how to analyze work methods and workstations, identify and resolve ergonomic problems, establish priorities, and implement an ergonomics program. It is a "reader friendly" guide.

The National Safety Council acknowledges with grateful appreciation the reviews and revisions of the following professionals in ergonomics: Thomas Armstrong, Kenneth A. Drew, Christopher Dockery, Dieter Jahns, Marilyn Joyce, Gary Lovested, Babak Naderi, Richard Niemeier, Donald Olsen, Larry Reed, Steven E. Snelling, and Denise Yanko. Special thanks also to Council staff, who devoted many hours to reviewing the manuscript in various stages: Barbra Jean Dembski and Jill Niland.

1 Ergonomics in the Workplace

MAKING ERGONOMIC IMPROVEMENTS in the workplace can have many benefits. Ergonomics is an engineering tool used to design-out job hazards, thereby eliminating the causes and preventing accidents. This is a very positive effect—designing-out the problem. Applying ergonomic principles to job analysis provides an opportunity to show employees that the company cares about preventing occupational injuries and illnesses and is serious in wanting to do something about them. Through the reduction or elimination of ergonomically related problems such as low back overexertion injuries or cumulative trauma disorders, work efficiency is improved. In summary, improving the ergonomic environment in the workplace can result in:

- Higher morale
- Improved quality of work
- Greater efficiency
- Improved productivity
- Reduced absenteeism
- Lower turnover
- Fewer lower back injuries
- Fewer repetitive trauma injuries.

Making ergonomic improvements to the workplace need not be expensive or overly complicated. Significant changes can be made if problems are approached with objectivity, good analysis of work methods, and strong knowledge of basic ergonomic principles.

This manual is designed to provide you with information and ideas that will help achieve those changes. Basic principles of ergonomics will be discussed, and you will be shown how to analyze work methods and workplaces. You will learn how to identify and resolve ergonomic problems and establish priorities.

This manual is intended as a quick-reference source to help you identify and correct ergonomics problems and to assist you in establishing and/or maintaining an ergonomics program. See the Bibliography at the end of this manual for other resources.

WHAT IS ERGONOMICS?

The word "ergonomics" comes from two Greek words: *ergon*, meaning "work," and *nomos*, meaning "laws." Ergonomics literally means, "laws of work." Ergonomics is an applied scientific/engineering discipline concerned with the interaction among systems and the people who operate and maintain them. Thus, whenever a human becomes part of the operation of a system, ergonomics becomes a necessary consideration. A system (such as a nuclear power plant for electric power generation), in most general terms, can be defined as a set of related elements organized to accomplish a given goal.

The Board of Certification in Professional Ergonomics, established in 1990, defines ergonomics as a body of knowledge about human abilities, human limitations, and other human characteristics that are relevant to design. Ergonomic design is the application of this body of knowledge to the design of tools, machines, systems, tasks, jobs, and environments for safe, comfortable, and effective human use.

The Occupational Safety and Health Administration (OSHA) defines ergonomics as the study of the design of requirements of work in relation to the physical and psychological capabilities and limitations of people; that is, ergonomics seeks to fit the job to the person rather than the person to the job. The aim of the discipline is to prevent the development of occupational disorders and to reduce the potential for fatigue, error, or unsafe acts through the evaluation and design of facilities, environments, jobs, tasks , tools, equipment, processes, and training methods to match the capabilities of specific workers. (*Federal Register,* vol. 57, no. 149, August 3, 1992.)

The field of ergonomics is very broad, and focuses on accommodating the information-processing capabilities of the brain and the physical properties and constraints of the body, as well as on designing controls to minimize muscle fatigue. While the modern, global term for this professional discipline is "ergonomics," in the United States, the term "human factors engineering" has also been popular since the 1940s. Figure 1–1 describes the views on ergonomics.

Ergonomics consists of several disciplines: engineering, physiology, medicine, anthropometrics, and behavioral science. All should be considered when establishing an ergonomics program. Other considerations should include:

- Psychological factors—attitude, motivation
- Somatic factors—age, sex, health, size
- Physiological factors—fuel (intake, storage), oxygen (pulmonary, cardiac)
- Learning factors—training, knowledge, adaptability

- Nature of work—intensity, duration, rhythm, technique, position
- Environment—heat, cold, noise, altitude, pollution.

These factors will be discussed in later chapters of this manual.

Ergonomics concentrates on the design of the system so that it is compatible with human capabilities and limitations. Thus, the objective of ergonomics is to design a system wherein all elements of the work environment promote the optimal effectiveness of the worker.

The ergonomist uses an approach that involves the systematic application of relevant information about human characteristics and behavior to the design and evaluation of equipment, facilities, and environments people use. The systems approach requires the ergonomist to be concerned with such factors as personnel selection, training and training methodologies, operational methods and procedures, and hardware and software design, to name a few. (See the Glossary for OSHA definitions of "systems approach" and other terms.)

This systems approach can theoretically be applied at two different stages, producing two types of ergonomics:

1. **Reactive ergonomics.** Corrective action is taken after an event occurs. For example, design and layout errors are corrected after an occupational injury or illness has occurred.

2. **Proactive ergonomics.** This is true accident prevention because a corrective action is taken *before* an occupational injury or illness occurs. For example, equipment design, work layout, and the work environment are analyzed for work hazards, and foreseeable hazards are eliminated before anyone is harmed.

Of course, the goal is that all corrective actions be proactive.

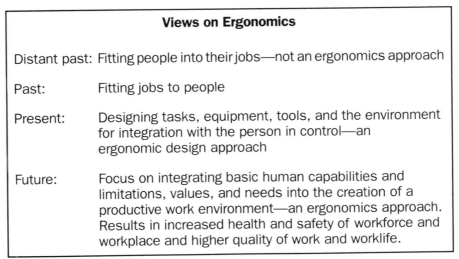

Views on Ergonomics

Distant past: Fitting people into their jobs—not an ergonomics approach

Past: Fitting jobs to people

Present: Designing tasks, equipment, tools, and the environment for integration with the person in control—an ergonomic design approach

Future: Focus on integrating basic human capabilities and limitations, values, and needs into the creation of a productive work environment—an ergonomics approach. Results in increased health and safety of workforce and workplace and higher quality of work and worklife.

Figure 1–1.

The Role of Ergonomics

The interaction between humans and the work environment covers a multitude of areas. The problem in organizing an ergonomics program is being able to ask the right questions and to define general problem areas that an ergonomics program should address. Figure 1–2 presents some of the major questions and problems that need to be considered.

Figure 1–3 lists some of the workstation and environmental issues. While this list is in no way complete, it does illustrate many possible causes of ergonomic problems. These ergonomic problems can lead to illnesses or injuries such as lower back overexertion injuries, cumulative trauma disorders, and other ergonomic-related accidents. To reduce such illnesses/injuries, the cause(s) must be identified and corrected through an understanding of the ergonomic principles that may be involved with the problem.

The contribution of ergonomics to an organization is broad and extends beyond reducing occupational injuries and illnesses. Figure 1–4 illustrates the impact of ergonomics in terms of benefits and manufacturing criteria that can be used to measure this impact.

Applying Ergonomic Principles

The goal of ergonomics is to make the workplace as adaptable as possible to the people who will be working in it. No two people are built

---

Basic Ergonomic Questions and Problems

Questions to Answer
- Where do we start?
- How do we form an ergonomics committee?
- What are the rewards?
- How do we evaluate our effectiveness?
- What problems do we solve?

Problems to Solve
- Tool/product design
- Workplace design
- Integration of new technology
- Environmental conditions
- Materials handling
- Task/job design
- Workstation design

Figure 1– 2.

Ergonomic Issues

Workstations
- Control and display design
- Location and orientation of work surfaces
- Posture/work height and reaches
- Movements/repetitions
- Vibration
- Accessibility
- Labels and location aids
- Guarding and warnings
- Hand tools
- Machinery and equipment design
- Size and shape of work objects
- Weight of work equipment and objects

Environmental
- Lighting/noise
- Temperature/humidity
- Housekeeping
- Hazardous materials.

Figure 1–3.

alike. Two people the same height may have a different range of arm reach. Two people the same weight may differ greatly in strength. A good workplace should accommodate at least 90% of the population. (See Anthropometry in Chapter 4.) One of the ways that this is done is by accommodating the physical characteristics of the person doing the job. The elements in the workplace or the individual job task that produce physical stress are first identified, and then altered to reduce the stress.

Each job has its own set of actions, and each job has its own level of stress. The amount of that physical stress is determined by several factors, including:

1. Force or weight handled
2. Load location or posture
3. Task repetition
4. Duration of task.

Physical stress can result from a poorly designed workplace. For example, a repetitive task done on a work surface that is too high may create stress to the shoulders. The nature of the job itself can also cause physical stress—a job that must be done in an extremely hot or cold environment, for example. It is important to remember that individual workers will react differently to a particular physical stress. A cold environment that is very stressful for one worker might be acceptable for another.

The following questions should give you some idea of what might cause physical stress in a given job:

- Does the worker sit or stand to do the job? Does the job require both sitting and standing? (See Combined Seated and Standing Work in Chapter 5.)

- Is the worker stationary while doing the job, or can he or she move about? (See page 40 for a discussion of dynamic versus static muscle activity.)

- Does the job require a great deal of strength or effort?

- Can any necessary reaching be done comfortably by most workers, or must they stretch and bend excessively?

- Is the work height comfortable, or must the worker perform the job in awkward positions (bent over a work table that is too low, for example)?

- Are job tasks extremely repetitive?

- Does the worker have any control over the pace of the job?

- Is the worker using the properly fitted tools?

- Is the proper personal protective equipment (PPE) used?

- Are there elements in the job environment that may cause discomfort (extreme temperature, contaminants, improper lighting, etc.)?

- Is the job boring to the worker?

You might find these elements combined in various ways. For example, a worker in a dusty environment might also be forced to do excessive reaching. These factors will be discussed in more detail in Chapter 3. This manual will deal with those factors as they apply to sitting and standing work, manual materials handling, cumulative trauma disorders of the upper extremities, and work with hand tools.

Before the workplace can be altered to reduce physical stress, the physical characteristics of the worker must also be determined. That is done by applying a science called anthropometry.

Anthropometry is the study of the physical dimensions of the various parts of the human body (i.e., size, breadth, girth, range of motion). These dimensions are expressed in terms of population percentiles (e.g., the 5th percentile male has an elbow-to-finger tip length of 17.6 inches—shorter than that of 95% of the male population). These dimensions are used in designing work tasks, workstations, and products that accommodate the individual worker or user.

Once you have the anthropometric data, it must be used correctly.

Contributions of Ergonomics	
Benefits	**Manufacturing Measures**
• Increased operator acceptance	• Injury/illness frequency and associated costs • Worker's compensation • Employee turnover • Employee absenteeism • Safety awareness
• Increased dependability or reliability	• Scrap (error) rate • Maintenance costs • Machine up-time • Logistic support • Safety awareness
• Flexibility and adaptability to change	• Reprogramming time/cost • Modifiability • Part change • Safety awareness
• Increased efficiency or productivity.	• Parts/hours (shift) • Scrap (error) rate • Safety awareness

Figure 1–4.

You have to consider both to whom you are applying the data and the source of the data. For example, you need to determine effective work heights for handling large, heavy parts by a predominantly male workforce. A study of a predominantly female workforce doing light assembly would not give you the appropriate information. If anthropometric data is not applied to the design of the work environment, physical stress can result. A further discussion of anthropometry and tables of some anthropometric measurements are presented in Chapter 4.

CERTIFICATION IN PROFESSIONAL ERGONOMICS

Prior to 1992, individual ergonomists or human factors professionals could be credentialed only indirectly through various types of state licensures for some of the professional fields related to ergonomics, such as engineering or psychology. The Board of Certification in Professional Ergonomics (BCPE) was established in 1990 as a nonprofit corporation to provide a formal organization and procedures for examining and certifying qualified practitioners of ergonomics. Because the Board regards ergonomics as synonymous with human factors and ergonomic design as synonymous with human factors engineering, qualified applicants may choose to be certified as either Certified Professional Ergonomists (CPE) or as Certified Human Factors Professionals (CHFP).

The Board certifies practitioners of ergonomics, not ergonomics researchers or theoreticians. A practitioner of ergonomics is a person who has (1) a superior knowledge of available ergonomics information; (2) a command of the methodologies used by ergonomists in applying that knowledge to the design of a product, system, job, or environment; and (3) a capability to apply his or her knowledge in the analysis, design, testing, and evaluation of products, systems, and environments. To qualify as an ergonomist, an individual must have at least a master's degree in ergonomics or a similar field and seven years of actual work experience. Certification is offered to ergonomists on an international basis.

2 Identifying Ergonomic Problems

Any attempt at ergonomic improvement should start with identifying where the ergonomic problems lie. The first step in the problem identification stage is a paper review of, for example, the Occupational Safety and Health Administration (OSHA) 200 Log form, internal accident investigations and medical reports, employee complaints, and so on. Walk-through surveys are also useful in identifying problems. Figure 2–1 lists factors to consider during a walk-through survey.

IDENTIFYING PROBLEM AREAS

There are two good methods of identifying problem job areas. The first is to analyze input from management-labor groups that have been set up to identify problem areas. The second method is to analyze data from such sources as worker's compensation files, medical reports, and OSHA or other governmental reporting forms.

When management does not require detailed justification for a project, the first method is a good one, but it is subjective and can vary in effectiveness. The second approach is more objective.

The data analysis approach takes into account the frequency and severity of OSHA-reportable injuries, summations of incidents, days lost by job, and costs of insurance and/or worker's compensation. This approach can be separated into two phases: record review and worksite risk analysis.

Record Review

Analyze injury and illness records (e.g., OSHA 200 Log) for evidence of cumulative trauma disorders and back overexertion injuries. Identify and analyze any trends and patterns found that relate to job classifications, departments, processes, or other work categories. Conduct surveys to identify those factors that may require a further comprehensive job study.

Calculate incidence rates per 200,000 hours worked for upper extremity disorders and back injuries, and compare them to the firm's overall incidence rate. Conduct surveys such as employee symptom surveys to identify jobs, departments, work process, and so on, that may require a further comprehensive job study.

A record review should highlight elements in the workplace that might contribute to ergonomic problems. There are many ways to analyze data, but the goal should always be to pinpoint problem areas and then set priorities for studying the jobs.

Worksite Risk Analysis

Ergonomics programs should contain provisions for a worksite risk analysis after the problem areas have been identified. A risk analysis identifies the extent of risk factors on the job. At this point, job risk factors are identified, quantified, and evaluated. This should be performed by qualified persons.

Often, jobs are videotaped so they can be played back, and risk factors are carefully identified and quantified to determine the real causes of the ergonomic problem (i.e., repetitively bending the wrist using an air tool, or repetitively lifting loads while the upper body is bent over 90 degrees).

The study should highlight elements in the workplace that might contribute to accidents, so that ergonomic problem areas can be pinpointed. These areas may involve both administrative and engineering functions.

Here are some useful categories for analysis:

- Department
- Job classification
- Shift or time of day
- Type of injury
- Body part
- Type of accident
- Instrument of injury (tool bin, machine tool, product, etc.)
- Job experience
- Day of week
- Age of worker
- Sex of worker
- Supervisor
- Machine name or number.

A matrix analysis method will permit the cross-reference of this information against OSHA Logs (No. 200 forms) or insurance data. This is time consuming, but the results are worth it. This method will place a cost or lost-time figure next to a suspected loss-producing or accident-causing element.

**Major Ergonomic Factors to Consider during a
Walk-Through Survey**

- Workstation and workplace design features
- Seating
- Postural demands
- Physical demands—manual handling, force, and energy requirements
- Perceptual load
- Display control and panel design features
- Machinery and equipment design features
- Information processing load
- Hand tool and implement design features
- Work organization and job design
- Work environment and climate
- Paced and shift work demands
- Mechanical hazards
- Housekeeping and maintenance
- Training

Figure 2–1.

There are many commercially available software packages that make it possible to do this type of analysis on a personal computer with much less expenditure of time. There are also many ways to analyze data, but the goal should always be to emphasize areas where ergonomic control will provide the most cost benefits. (See Chapter 3.)

INDICATORS OF ERGONOMIC PROBLEMS

Work accidents can indicate an ergonomic problem, but there are other, less obvious indicators, including:

- Apparent trends in accidents and injuries
- Incidence of cumulative trauma disorders
- Absenteeism, high turnover rate, temporary or seasonal hiring patterns
- Employee complaints
- Employee-generated changes in the workplace (e.g., tool modification)
- Incentive pay systems
- Excessive overtime and increased work rate
- Poor product quality
- Manual materials handling and repetitive motion tasks
- Improperly designed workstations for those with disabilities or reduced capabilities.

These are not the only situations that indicate there may be an ergonomic problem, but they are the most common in a typical manufacturing environment. These indicators will be examined in more detail.

Apparent Trends in Accidents and Incidents

Problems relating to manual materials handling or cumulative trauma disorders can be detected by analyzing incidence rates and employee injury/illness trends. Examine the OSHA 200 Logs (Log and Summary of Occupational Injuries and Illnesses), or other governmental forms, nursing first aid logs, and insurance carrier claims loss runs. These and other internal reports should be reviewed periodically.

Incidence of Cumulative Trauma Disorders

Employee exposure to cumulative trauma disorders should be determined. These disorders are recorded under column 7(f) on the OSHA 200 Log. A survey of incidence records, examination of the workplace for potential hazardous exposure, and a detailed job analysis should provide the information you need.

Some of the more common cumulative trauma disorders are tendinitis, carpal tunnel syndrome, and epicondylitis (tennis elbow). Some cumulative trauma disorders develop only after long-term exposure and forceful movements, so incidence data alone may not reveal the true pattern of exposure. You will also need to analyze incidence of illness, since that is how cumulative trauma disorders must be reported.

Absenteeism, High Turnover Rate, Seasonal Hiring Patterns

Job conditions that create excessive physical or mental stress may result in absenteeism and high job turnover. Monotonous jobs with little or no challenge are also stressful, and may also result in higher than normal absenteeism and turnover. Both types of jobs should be assessed ergonomically.

In some industries, workers are hired for short periods to do jobs that require minimal training and involve repetitive assembly tasks. These short-term tasks can expose such workers—who are unaccustomed to the activities involved—to various types of cumulative trauma disorders. If it is not practical to institute engineering changes to solve these problems, other solutions can be found. (See Chapters 7 and 8 for discussion of such solutions as job rotation or gradual rotation from less- to more-demanding activities.)

Employee Complaints

If workers are complaining, listen to what they have to say. They can provide valuable insight into ergonomic problems in the workplace, as well as practical ideas for solutions.

Employee-Generated Changes in the Workplace

Awareness of what your workers are doing can help you identify the need for an ergonomics program. Look for changes in the workplace made by the workers themselves—especially after new equipment or production processes have been installed. That is a good indication the workers are having problems with the job. Here are some typical signs to watch for:

- Workers have added padding to hand tools, or have put padding on the edges of equipment or work surfaces.
- Seats and chairs have been modified.
- Makeshift platforms are being used to stand on.
- Outside help is being used to get the job done.
- Workers have designed or modified their own personal protective equipment.
- Workers have made their own changes in the design of the work flow.
- Illumination (or ventilation) has been modified.
- Tools, fixtures, or staging areas have been realigned.

If workers seem to be taking more frequent breaks or rotating jobs more often, that might indicate a potential problem area. Surfaces at a workstation that have been rubbed shiny by continual contact with workers' bodies are another sign of potential trouble.

Each of these signs is a warning or symptom that something is causing stress in a particular job or work environment. That kind of stress promotes injuries/illnesses and decreased productivity.

Incentive Pay Systems

Incentive pay based on job-rate quotas can give workers higher income and enhance the company's productivity. However, without adequate ergonomic controls, that kind of system could contribute to increased rates of disorders.

For example, workers who increase their work rate may overstress their bodies. A worker might ignore the early symptoms of cumulative trauma disorder, and continue working until an injury/illness occurs. If caught early, cumulative trauma disorders can be treated conservatively and effectively. If treatment is delayed, the employee can lose work time and possibly even require surgery.

The problems associated with incentive pay systems may be eased by job enlargement, employee rotation, and by including fatigue allowances in the job procedures.

Excessive Overtime and Increased Work Rate

If workers are putting in many hours of overtime, or if their work rate has been increased, fatigue can affect job performance. The fatigue is caused by the increased demands on a worker's physical and physiological energy (see Chapter 4). When a worker is fatigued, there is greater potential for job errors and accidents. The ability to make some rational decisions may also be impaired. When these conditions exist, an ergonomic assessment should be made.

Poor Product Quality

If product quality is poor, or if quality control measures seem to be inadequate, ergonomic intervention might be needed. Customer complaints, audits of a product that has already been inspected, and the need for job-sample tests are all signs that ergonomic problems may exist.

The point where a product is actually made is a potential site of ergonomic problems, but it is not the only one. Poor ergonomic conditions in the quality control area can affect inspection performance. This could result in a substandard product passing inspection, or an acceptable product being rejected. Conditions that can influence the performance of quality-control inspectors include workstation design, work flow design, pace, lighting, noise levels, and background colors.

Manual Materials Handling and Repetitive Motion Tasks

Requiring workers to lift, lower, push, pull, or carry items can result in back pain or injury. The National Institute for Occupational Safety and Health (NIOSH) concluded in 1981 that there was significant increase in musculoskeletal injuries when one or more of the following four factors were present:

- The object lifted was large or difficult to handle.
- The object lifted was bulky.
- The lift originated from the floor.
- The frequency of lifting was high.

NIOSH classified the factors that can influence the safety of a lifting task into two broad categories: job risk factors and personal risk factors.

The job risk factors that can influence the probability of injury while performing manual materials handling tasks are:

- **Weight**—What is the weight of the object, and how much force is required to move it?
- **Location**—How far is the load from the worker in terms of both horizontal and vertical distance at the beginning of the lift?
- **Frequency/Duration**—How often is the task repeated during a job cycle and during the work shift?
- **Stability**—Is the load bulky or compact with respect to its center of gravity?
- **Grip**—How easy or difficult is it to hold on to the load? If the load has handles, what is the handle texture, size, and shape?
- **Workplace geometry**—How does the design of the workplace affect lift distances, posture, and the necessity to twist the torso while performing the task?
- **Environment**—How warm or cold is it? Is there noise, vibration, or friction?

The personal risk factors that can influence the probability of injury while performing manual materials handling tasks are:

- **Sex**—The average female has approximately 65% of the lifting strength of the average male.
- **Strength**—Not all people of the same sex have equal musculoskeletal strength, and this affects their ability to perform lifting tasks.
- **Age**—Muscle strength diminishes slowly with age. At age 55, the average person has lost approximately 15% of the muscle strength he or she had at age 25.
- **Fitness**—Regardless of a person's strength, his or her potential for injury is affected by the degree of flexibility, endurance, and total body muscle tone.
- **Anthropometry** (size and proportions of the body)—Differences in body weight and stature can affect lifting abilities.

- **Lifting techniques and training**—Workers can be taught lifting techniques that will provide biomechanical advantages.

Workstations of People with Disabilities

A disabled person is one who has a physical or mental impairment that limits a major life activity. The Americans with Disabilities Act requires that feasible ergonomic accommodation for accessibility and adaptation to an individual's particular disability be made to the workstation and surrounding areas. That may require design or purchase of supportive aids or devices.

WHEN TO IDENTIFY ERGONOMIC NEEDS

The earlier ergonomic needs are dealt with, the better. That means these needs should be considered during the design phase for new processes, and at the point when new equipment purchases are first contemplated. Changes made after a new process has been implemented, or a new piece of equipment installed, usually waste time and money. In most cases, the results will also be less than satisfactory.

New equipment and processes that are put in place without taking ergonomic considerations into account may cause problems. Some problems might appear immediately, but others might not surface for a long time. This is especially true of cumulative trauma disorders of the hand and wrist. They are illnesses that result from long-term exposure to repetitive tasks done in awkward postures or using excessive force.

Never assume that because equipment or processes are new and up-to-date, they will be ergonomically correct. Facility engineers may not consider ergonomic factors for the areas in which a worker interfaces with a machine. Ergonomic principles and the human element in the workplace tend to be minimized in the usual industrial or mechanical engineering curriculum.

In general, it is a good policy to review the health and safety aspects of all new manufacturing processes, design, and major equipment purchases. Purchasing, engineering, and maintenance departments, and tool room staff should help review manufacturing processes, design, and major equipment. When a new production line or equipment purchase is being considered, use these factors to evaluate the design:

- Physical capabilities of the worker
- Handling methods between operations
- Actual motions made while doing the job

Ergonomic Principles

Workplace Design and Layout

Head height consideration:
- Allow for the tallest
- Natural posture is to look down slightly
- Avoid narrow viewing angles

Shoulder height consideration:
- Place items and parts between shoulder and waist height
- Avoid reaches above the shoulder

Elbow height consideration:
- Normal work—raise surface to just below elbow height
- Very precise work—raise the surface above elbow height and provide rests for arms
- Heavy work—place work surface about 8 inches below elbow

Arm reach consideration:
- Allow for the shortest when reaching up or out
- Allow for the tallest when reaching down
- Keep frequent work within forearm distance

Knuckle height consideration:
- Keep lifting tasks between knuckle and shoulder height

Leg length consideration:
- Allow for long legs
- Provide adjustments or footrests for shorter legs

Hand size consideration:
- Allow for small hands for grasping
- Allow for small hands for dangerous openings
- Allow for large hands for access

Body bulk consideration:
- Allow for the largest
- Remember that clothing adds bulk

Other items to check for:
- Horizontal reach
- Vertical reach
- Leg room, twisting, upper body posture
- Proper workplace design and layout reduces unnecessary stress

Factors in Design Guidelines for Adjustability

A good fit between the person and the task can be obtained by making the workplace adjustable:
- Raise or lower work surface
- Change the worker's position
- Use a tool to extend a reach
- Adjust shape
- Adjust location
- Adjust orientation
- Adjust chairs
- Use support stools, swing bracket stools, and other props
- Use platforms, step-ups, and mechanical lifts

- Use footrests
- Use armrests
- Adjust design and location of tools
- Adjust trays and carriers

Adjusting the product or product containers:
- Relocate or reorient product or work object
- Use jigs, clamps, and vises
- Use parts bins
- Use lift tables, elevators, and similar equipment

Part Manipulation Considerations

Range of movement:
- Keep well within full range of movement
- Never hyperextend any joint
- Keep elbow at right angles when applying force
- Keep wrists straight

Rate of repetition:
- Allow frequent job rotation
- Allow self-pacing (new employees should start off at a slower rate)
- Keep rates to a minimum
- Allow frequent rest pauses
- Check fatigue in specific muscles

Length of contraction:
- Keep elbows to the side
- Keep static muscular contractions (tensed muscles) to a minimum
- Minimize time of muscular contraction
- Minimize force of muscular contraction

Force of application:
- Never exceed 20% of operator's maximum force in a prolonged or repetitive work posture; avoid frequent impulses
- Remember that legs, knees, hips, and shoulders may all be subjected to similar stresses

Figure 2–2.

- Tools required for the job
- Dimensions of the physical work environment, including such things as work heights and reaches the worker must make while sitting or standing
- Environmental issues, such as equipment noise, amount of heat released, illumination needed.

See Figure 2–2 for a more detailed list of ergonomic principles to consider.

Taking those factors into consideration, the workplace and the equipment should be designed to provide efficient, accurate, and safe operation. Jobs should be designed to promote correct body posture. Building adjustable features into the workplace will greatly increase the range of workers able to perform a given task. It should be possible to adapt the workstation and the work environment to the capabilities (both physical and mental) of the worker.

In addition to evaluating the design of new equipment, consider auxiliary equipment that might be used with it, such as:

- Chair design
- Height and location of stock containers
- Hand tools used on the job
- Mechanical aids that might be needed.

EVALUATING THE FACILITY

See Appendix 1 at the end of this manual for a checklist that will aid in evaluating the overall facility. This evaluation will uncover problem areas that can be examined further and help you develop priorities to perform a more detailed analysis. Any "yes" answers indicate that something may be wrong with the job.

3 Ergonomic Task Analysis

AFTER PROBLEM JOBS HAVE BEEN IDENTIFIED, the next step is to identify the cause of an ergonomic problem. This may require a simple ergonomic task analysis or a detailed one, depending upon the complexity of the problem. This is likely to be a rewarding experience, and one in which something new is learned almost every time. In addition, an in-depth ergonomics study is an excellent way to get to know a company's manufacturing practices, engineering methods, and the people who perform the jobs.

Before a task analysis program is started, establish good lines of communication with everyone involved. This will reduce the natural tendency of workers to be suspicious of someone who suddenly arrives at the workstation with cameras, notebooks, and a lot of questions. Introducing the program properly can also give you the benefit of the workers' own insights.

Before the task analysis program is started, these questions should be answered for everyone involved:

- What are the goals and intent of the ergonomic analysis?
- Who will be involved?
- Exactly what is going to happen?

Decide on the best way to inform the people at your workplace. Small group meetings, bulletin board notices, company newspapers, and memos are all good ways to get the information out to people.

Task analysis is a method of studying the employee in the overall work environment. Properly executed, it will show you the restrictions imposed on the workers and the inefficiencies of the workstation.

Task analysis can pinpoint the following problems or needs:

- Work tasks that involve adverse movements
- Tasks involving excessive manual lifting
- Wasted motion or energy
- Poor operational flow
- Work tasks that do not present the potential for psychological stress
- Fatigue factors
- Need for automation
- Quality control problems.

PERFORMING THE ERGONOMIC TASK ANALYSIS

The goal of job task analysis is to collect data that can be used to identify the causes of ergonomics-related problems. The data can be collected in one or more of the following ways:

- General observations
- Questionnaires
- Interviews
- Video analysis
- Photography
- Drawing or sketching
- Evaluation of worker capabilities
- Measurement of various risk factors (such as repetition, force, posture, etc.) related to ergonomic problems.

General Observations

Direct observations of the worker, work tasks, and the work environment are common and useful methods of obtaining job analysis data. Direct observation is particularly useful for studying jobs with low-level skill requirements and jobs with short, repetitive cycles. These jobs can also be studied by interviewing the workers or actually performing the task involved.

Look for work tasks or situations that produce repetitive trauma disorders or injuries related to manual materials handling. (See also Chapters 6 and 7.) Some specific examples are:

Repetitive Trauma Disorders
- Awkward postures
- High task repetition
- High force required
- Mechanical stress points
- Cold temperatures
- Fit of work gloves
- Poor workstation design
- Hand tool fit
- Vibration sources.

Manual Materials Handling
- Lifting from the floor
- Lifting while twisting
- Lifting heavy weights
- Lifting bulky objects
- Lifting repeatedly
- Lifting above shoulder height
- Lifting heavy items while seated
- Pushing or pulling loads
- Nature of hand holds
- Storage methods
- Parts staging
- Floor condition in the work environment
- Workstation layout
- Conflicting movements
- Poor posture
- Bending.

Whether or not any of these tasks or situations found on the job could result in injury depends upon numerous factors, such as the task performed, the weights and forces handled, and the number of times the task must be repeated during a work shift. (See Chapters 6 and 7 for a more detailed discussion.)

Appendix 2 at the end of this manual is a workstation checklist that can be used to observe job performance. Each question refers to an area that could need attention. The items in the checklist range from broad, general areas to extremely specific details. This range will allow for various levels of analysis of the job tasks under observation.

Some of the problems identified by this checklist can be corrected by the maintenance department, while others might be taken under consideration by the personnel department as a factor in employee placement. Others might be noted for long-term corrective action.

Once the job has been observed, it is a good idea to get the worker's input. The worker should know the job better than anyone.

Questionnaires and Interviews

Workers can be surveyed by questionnaire or interview. Questionnaires will make it easier to tabulate results, but you may not get the whole picture without talking to the workers. The person analyzing the data should be able to judge the best approach for a given workplace.

If questionnaires are used, it is useful for the analyst to guide the worker through the questions to be sure the answers are consistent with the intent of the questionnaire. That approach will give maximum validity to the results.

Appendix 3 is a sample of the kind of questionnaire that can be used to guide employee interviews. It is designed to deal with a very specific problem.

Another type is the employee symptoms questionnaire in which employees mark an 'X' on the body part on a form indicating an area of discomfort (Figure 3–1). At the same time the employee is asked to rate the level of pain (e.g., a rating scale of one to five, with five being high pain or discomfort). This type of questionnaire can be easily administered by a supervisor, and responses can be quickly compiled. Consistent negative responses on a questionnaire may indicate a pattern of ergonomic problems that needs further investigation.

As with any questionnaire or interview, it is recommended that the results be shared with the employees. If the information from the survey does not lead to some type of corrective activity, chances are employees will be discouraged from participating or providing good responses in any future surveys.

Virtually every area observed using the checklists can be discussed directly with employees. Decide what your priority areas are, and ask specific questions about them.

Video Analysis

Videotape is a good tool for collecting and analyzing task data. Videotapes can be reviewed at a later time, and trouble spots can be viewed repeatedly or in slow motion or "stop motion" to allow detailed analysis. Video cameras that provide internal stopwatches and character generators to print titles and lablels can also be useful for timing and labeling an analysis.

When selecting a video camera and recorder/playback unit for use in ergonomic analysis, try to find one that can be used in low light. It is best if the camera is lightweight and portable, and it should provide sound recording.

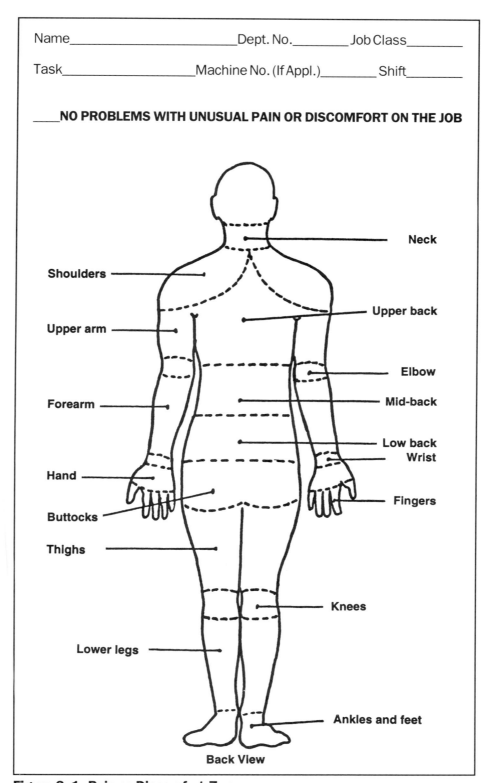

Name_____ Dept. No._____ Job Class_____

Task_____Machine No. (If Appl.)_____ Shift_____

____**NO PROBLEMS WITH UNUSUAL PAIN OR DISCOMFORT ON THE JOB**

Neck

Shoulders

Upper back

Upper arm

Elbow

Forearm

Mid-back

Low back
Wrist

Hand

Buttocks

Fingers

Thighs

Knees

Lower legs

Ankles and feet

Back View

Figure 3–1. Pain or Discomfort Zones

Photography

Video, 35mm photographs, and slides are all good for analyzing a job situation. Videotaping is an easy way to capture all of the postures and tasks in motion. These can be analyzed (and timed) to identify specific risk factors by the task causing the problem.

Drawing or Sketching

A simple drawing of the work area and work process might be all you need to work with for later evaluation of the work process. To be sure a work process and workstation are represented completely, ask an employee to do the drawing, or at least check over the finished drawing to be sure it is correct.

Evaluation of Worker Capabilities

Measurements are an important part of workplace and workstation ergonomic analysis. Measurements, such as length of reach, work heights, task frequency, and energy expended all describe a task. To say that a particular task involves objects that are "too heavy to lift" or "too hard to push" depends upon the extent of these task measurements. Measurements should be done by knowledgeable, trained individuals.

Biomechanic measurements are needed to do calculations such as the NIOSH Lifting Task Analysis. Many think of worker strength as the main criterion for work. However, in jobs requiring excessive repetitive tasks, the energy expended on the job may be the limiting factor, not the worker's strength. Energy expenditure can be measured, but it is not usually done in the work environment because it requires complex equipment and special skills. In general, people can work for long periods of time without undue fatigue if they work at about one-third of their maximum capacity. Chapter 6 will discuss the specific task measurements in more detail, including how to measure and evaluate their importance. Figure 6–5 shows some of the task variables to measure.

ANALYZING DATA AND RECOMMENDING CHANGES

After task measurements have been taken (e.g., weights and forces handled, frequency of tasks) and other parameters noted (e.g., employee suggestions, environmental observations), assemble the information and evaluate it to determine if an ergonomic problem exists, and if so, to what degree. After the information is evaluated and specific causes are identified, the next step is to correct the problem.

4 Anthropometry and Workplace Stressors

WORKPLACE STRESS can be induced by physical, physiological, biomechanical, and psychological aspects of the worker/workplace interaction. A workplace that takes into account only the physical dimensions and capabilities of individual workers is likely to be only partially ergonomically correct. The effects of fatigue and stress on psychological and mental capabilities should also be considered.

ANTHROPOMETRY

Anthropometry literally means "measuring the human." Application of anthropometry in the workplace is one of the essential tasks in developing an effective ergonomics program. The human body is traditionally measured in terms of height, breadth, depth, and distance. All of these are straight-line, point-to-point measurements between landmarks on the body and/or reference surfaces. Curvatures and circumferences following body contours are also measured.

However, there is no such thing as a "typical" person. No worker is likely to be average in every physical dimension. Furthermore, the population of workers in your workplace is likely to be different from the population of workers in other workplaces. It is possible to design workstations so that 90% of the working population is accommodated ergonomically, but this means that the 5% tallest and 5% shortest individuals will not fit ergonomically in these workstations.

The anthropometric data in Tables 4–1 and 4–2 will help in designing workstations and job tasks that will accommodate a large percentage of workers in terms of work heights and reaches. The anthropometric data provided in these tables were adapted from the *1988 Anthropometric Survey of U.S. Army Personnel: Methods and Statistics*. While the data is biased toward the younger segment of the population, matched studies indicate a close correlation between civilian and military populations, except for heights and weights of female civilians.

Through good interpretive judgment, this data can be used for

Table 4–1. Selected Body Dimensions (without Shoes) in Inches for Males/Females Population Percentiles

Heights	Population Percentiles			
	5th pct'ile	50th pct'ile	95th pct'ile	Std. Dev.
Standing:				
Stature	64.8/60.2	69.1/64.1	74.4/68.4	2.63/2.51
Eye height	60.2/55.7	64.3/59.6	68.6/63.8	2.59/2.46
Shoulder (acromium)	52.8/48.8	56.8/52.4	60.8/56.4	2.44/2.28
Elbow to floor	39.2/36.5	42.2/39.2	45.4/45.3	1.89/1.76
Foot length	9.8/8.8	10.6/9.6	11.5/10.4	0.52/0.48
Weight (lb)	135.8/109.4	171.3/135.0	216.2/169.7	24.6/18.4
Sitting:				
Height (to seat)	33.6/31.3	36.0/33.5	38.3/35.8	1.40/1.37
Eye height (to seat)	28.9/27.0	31.2/29.1	33.4/31.3	1.35/1.31
Shoulder (to seat)	21.6/20.0	23.5/21.8	25.4/23.8	1.16/1.13
Knee top to floor	20.2/18.7	22.0/20.2	23.8/22.0	1.10/1.04
Popliteal*	15.5/13.8	17.1/15.3	18.8/16.9	0.98/0.93
Lengths				
Elbow to fingertip	17.6/16.0	19.0/17.4	20.6/19.0	0.92/0.92
Buttock to knee length (sitting)	22.4/21.3	24.2/23.1	26.3/25.2	1.18/1.17
Wrist to center of grasp	2.4/2.3	2.7/2.6	3.1/3.0	0.19/0.19
Functional grip reach (backward/ forward)	27.3/24.9	30.0/27.0	32.0/29.3	1.45/1.33
Elbow to center of grip	13.1/11.8	14.1/12.9	15.4/14.1	0.70/0.70

* The vertical distance from the floor surface to the back of the right knee crease when the subject is seated.

Adapted from *1988 Anthropometric Survey of U.S. Army Personnel: Methods and Statistics*, NATICK/ TR-89/044, U.S. Army, Natick RD&E Center, Natick, MA, 1989.

Table 4–2. Selected Body Dimensions (without Shoes) in Centimeters for Males/Females

Heights	Population Percentiles			
	5th pct'ile	50th pct'ile	95th pct'ile	Std. Dev.
Standing:				
Stature	164.7/152.8	175.5/162.7	186.6/173.7	6.68/6.36
Eye height	152.8/141.5	163.3/151.4	174.3/162.1	6.57/6.25
Shoulder (acromium)	134.2/124.1	144.2/133.2	154.6/143.2	6.20/5.79
Elbow to floor	99.5/92.6	107.1/99.7	115.3/107.4	4.81/4.48
Foot length	24.9/22.4	26.9/24.4	29.2/26.5	1.31/1.22
Weight (kg)	61.6/49.6	77.7/61.2	98.1/77.0	11.1/8.35
Sitting:				
Height (to seat)	85.4/79.5	91.4/85.1	97.2/91.0	3.56/3.49
Eye height (to seat)	73.5/68.5	79.2/73.8	84.8/79.4	3.42/3.32
Shoulder (to seat)	54.8/50.9	59.8/55.5	64.6/60.4	2.96/2.86
Knee top to floor	51.4/47.4	55.8/51.4	60.6/56.0	2.79/2.63
Popliteal*	39.5/35.1	43.3/38.9	47.6/42.9	2.49/2.37
Lengths				
Elbow to fingertip	44.8/40.6	48.3/44.2	52.4/48.2	2.33/2.34
Buttock to knee length (sitting)	56.9/54.2	61.5/58.8	66.7/64.0	2.99/2.96
Wrist to center of grasp	6.2/5.9	6.9/6.6	7.8/7.5	0.49/0.49
Functional grip reach (backward /forward)	69.3/63.2	75.0/68.5	81.3/74.4	3.68/3.39
Elbow to center of grip	33.2/30.0	35.9/32.8	39.1/35.8	1.79/1.77

* The vertical distance from the floor surface to the back of the right knee crease when the subject is seated.

Adapted from *1988 Anthropometric Survey of U.S. Army Personnel: Methods and Statistics*, NATICK/ TR-89/044, U.S. Army, Natick RD&E Center, Natick, MA, 1989.

civilian populations. Only a limited number of physical dimensions are presented. The reader is encouraged to consult the study noted at the bottom of the tables for additional dimensions and for more in-depth information.

Other factors might influence the way anthropometry affects design. These include elements specific to a given workplace, such as heavy clothing for work in cold areas or personal protective equipment.

If an existing workstation hinders a worker's ability to work comfortably by requiring extended reaches, twisting, or awkward posture, the workstation can be redesigned or modified.

Anthropometric data are available from several sources and should be consulted when designing or modifying a workstation or piece of equipment. It is beyond the scope of this manual to cover the subject in substantial detail, except to emphasize that the workstation should be designed to accommodate the varying sizes and reaches of workers. Because one fixed-dimension workplace design will seldom fit everybody in the workplace, adjustability of design is the key. For sources of anthropometric data and how to apply them to workplace design, see the Bibliography at the end of this manual.

STRESSORS

Stress results when outside forces exceed a tolerance level within a person, resulting in bodily or mental tension. Sources of stress may be physiological, biomechanical, or psychological in nature. Table 4–3 presents examples of both physiological (or environmental) and psychological stress. In many situations, it is a complex interaction of psychological, physiological, and biomechanical factors that induce stress.

Fatigue

There is no precise definition of fatigue. Sometimes the term is used to refer to general feelings of tiredness, sometimes it is meant to relate to a reduction in work output, and sometimes it refers to the physiological conditions resulting from continued work activity. One definition of fatigue is: A state resulting from boredom, from unreasonable expenditure of energy, from physiological changes in the body, or from undue biomechanical forces.

Manifestations of fatigue appear in many forms, including the following symptoms:

- Muscular soreness
- Aches
- Sleepiness
- Mental confusion
- Muscular tension
- General weariness.

In order to better understand the concept of fatigue, it is helpful to classify it into types. One way to distinguish types of fatigue is to identify fatigue as either subjective or objective. Subjective fatigue is present when there are feelings of tiredness. General weariness is enough evidence to infer subjective fatigue. Objective fatigue results when there is a decrement of energy-producing materials in the body (e.g., sugar) and when waste products (e.g., carbon dioxide and lactic acid) accumulate. Subjective and objective fatigue do not always appear hand-

Table 4–3. Sources of Fatigue and Stress

Environmental (Physiological/Biomechanical)	Psychological
Kinetic stress	Danger
Acoustic noise	Fear
Vibration	Monotony
Force	Torque
Thermal	Task speed
Atmospheric pressure	Task load
Chemical	Sensory deprivation
Radiation	Distractions
Work overload (physical)	Threatened failure
Glare	Uneventful vigilance
Sleep deprivation	Circadian rhythms

in-hand, because subjective stages of fatigue may often be reduced by specific changes in work conditions.

Though we usually think of fatigue as having primarily physical effects, one can consider the idea of mental fatigue. Mental fatigue is a tiredness that occurs from mental, rather than physical work. Difficult mental operations can be tiring, especially if one encounters any amount of frustration or conflict in his or her cognitive and conative processes. Mental fatigue is usually composed of general feelings of strain and effort. Boredom, drowsiness, and sleepiness may be concomitant symptoms.

Fatigue may also be described as either acute or chronic. Acute fatigue refers to the fatigue that develops during a working day. A simple example of acute fatigue is the discomfort produced by pacing about in an office or a job site. This type of fatigue is remedied with normal sleep and rest. The most noticeable effect of acute fatigue is a reduction of work productivity or a decrease in efficiency. Chronic fatigue is a condition characterized by low motivation and poor work productivity, yet is not affected by work cycles, sleep, or rest. It may also be a result of a medical or toxicological condition.

There is also cumulative fatigue, which is said to result after an extended series of work cycles. Usually the work cycles are interspersed with little opportunity for rest and recovery. Productivity and efficiency levels do not change much more than in the acute phase. However, workers note that more effort is demanded on their part to maintain proficiency levels, and they doubt their abilities to persevere. Emergency situations that require an operator, for example, to maintain extended work shifts set the stage for cumulative fatigue to appear. Under these circumstances a limited number of workers, operators, and engineers must attend to an emergency problem over a number of days. Extending full efforts at work and having little opportunity to rest usually results in some degree of cumulative fatigue.

Physiological Stressors

Kinetic stress results when the body is subjected to rapid accelerations as is the case in vehicular maneuvers. This brings about changes in the blood flow and may cause loss of peripheral vision, illusions of motion, loss of fine motor control, total loss of vision and, finally, unconsciousness. Such conditions may render an operator completely helpless. Vibrations and acoustic noise may be considered together because they are both mechanical oscillations. Oscillations in the range of 20–20,000 hertz are heard as sound. Oscillations below 20 hertz are perceived as vibrations. Vibration and noise may interfere with

Learning Resources Center
ASHEVILLE-BUNCOMBE TECHNICAL COMMUNITY C
17 NOV 99 10:04AM
Renewals

Patron Barcode:23312000197634

Renewal: Ergonomics at work : human fact
33312000575050   Due: 01 DEC 99 *

Renewal: Ergonomics : a practical guide.
33312000399519   Due: 01 DEC 99 *

"'Tis education forms the common mind,
    just as the twig is bent,
    the tree's inclined."
       -- Alexander Pope [17341]

other sensory functions and disrupt motor coordination and actual feedback. A person reacts to noise and vibration with an arousal pattern that serves as an alarm reaction. Persistent noise induces anxiety and irritation even at low levels of intensity. Severe forms of vibration and noise interfere with internal organs and the central nervous system; this can lead to disability, unconsciousness, and even death.

Thermal stress refers to that stress induced by heat and cold. Extreme heat may cause heat stroke. Symptoms of heat stroke are accelerated heart rate and cardiac arrhythmia. Factors that affect this may include temperature, humidity, energy expenditure, and metabolic functions. Humans can develop a tolerance to heat by acclimation and thereby reduce the effect of heat stress. Though exposure to cold does not affect sensory ability (with tactile abilities excepted), it does affect coordination and motor ability quite readily. Hand strength is reduced because of blood vessel constriction and reduced blood flow. The reduction in blood circulation through the extremities limits both motor strength and persistence of effort.

Atmospheric pressure stress occurs when humans subject themselves to different atmospheric conditions in sea and space explorations. By doing this, humans breathe gases of different compositions and abnormal pressures. Of course, this has biological effects. Physiological stress will result as a function of abnormal gaseous levels of components in the blood. Decompression sickness and anoxia are examples of reactions to changes in atmospheric conditions. The primary consideration is to avoid anoxia, the lack of oxygen. Variations of nitrogen, carbon dioxide, and other gases in blood levels also affect the regulation of many physiological processes.

Chemical stressors include airborne and ingested chemicals. Eye irritation is one effect of airborne chemicals. Smog is one cause of eye irritation, and although smog does not interfere with visual acuity, it may induce eye strain or fatigue. Ingested chemicals such as alcohol may also serve as stressors. Alcohol, for example, impedes motor control and degrades visual search and tracking capabilities.

Radiation stress occurs when a person is exposed to ionizing radiation, that part of the electromagnetic spectrum that contains X rays, gamma rays, and cosmic photons. Any exposure or perceived exposure can cause some people to suffer psychological stress, with symptoms of weakness, apathy, inactivity, and nausea. However, massive doses (500–1000 rems) of radiation are capable of destroying the nervous system. Convulsions occur and the victim goes into shock, with death occurring soon afterward. Lower levels of radiation are used in the treatment of cancer patients (about 200–300 rad). Reactions of these patients to radiation therapy appears in the form of nausea, headaches, and fatigue. Radiation exposure does not always result in performance decrements, but at high levels this is a certainty.

Physical work overload consists of muscular exertion and other excessive demands of the body. Extreme muscular exertion affects coordination and strength. Should a person continue to overexert himself or herself, oxygen and sugar levels in the blood will be drastically reduced. Muscle spasms and collapse are the eventual consequences of overexertion.

Glare is an example of an environmental stressor that affects visual performance. Glare is brightness that is substantially greater than normal levels of luminance in the visual field. This causes annoyance, discomfort, and degraded visual abilities. Glare does not allow optical contrast levels, and therefore affects visual detection abilities. It also results in eye strain and fatigue.

Sleep deprivation is a stressor that has many effects on performance. As noted earlier, sleep is necessary in the reduction of acute fatigue symptoms. Lack of sleep may result in weariness, decreased strength, irritation, motor coordination problems, poor visual abilities, and even in hallucinations. Decision-making ability is one cognitive function that suffers from lack of sleep. Mental operations become slower and more difficult if adequate rest is not obtained. Though there is great variability in an individual's ability to go without rest, no person is immune to the effects of lack of sleep.

Psychological Stressors

Psychological stressors are those sources of stress that cause mental tension. Because tension may ultimately result in physiological disturbances, it is sometimes impossible to differentiate between physiological and psychological stressors. Some special psychological stressors were listed previously in Table 4–3.

It is not the goal of the ergonomist to eliminate all sources of stress, because not all levels of stress are disruptive. Indeed, a limited amount of stress is desirable because it serves to motivate a worker. It is the concern for safety that, for example, predisposes a nuclear power plant operator to frequently scan annunciators and selected groupings of displays. An inordinate amount of concern or anxiety may cause the operator to spend too much time focusing on selected displays and hence neglect his or her other duties. This results in a decrease in overall performance. So it becomes obvious that stress can be either disruptive or facilitative, depending on its degree.

Fear and danger are known to have a stressful effect on humans because they put a person in a state of continuous alarm. Persistence of this state of alarm is wearisome, both mentally and physically. The functions of the autonomic nervous system are stepped up so that heart rate and blood pressure increase, and breathing becomes shallow and rapid. The mind maintains a vigilant state that allows little chance for rest and recovery from mental tension. By measuring the heart rates and chemical levels of people experiencing fear and danger, (e.g., combat soldiers), researchers have noted some of the physiological results of these stressors. Certain chemicals, such as adrenal cortico-steroids, are produced in greater quantities. These kinds of chemical imbalances may result in serious damage to the internal organs in just the same way that ulcers may be the result of chemical changes of psychic origin.

Fear and danger may also interfere with psychological processes. Attention and cognitive functioning may be disrupted. Persons may panic or perform nonadaptive behaviors because of fear. Extreme danger may predispose one to abandon his or her duties altogether. One common response to a dangerous situation is to deny its presence altogether. This observation provides some explanation, for example, for why some operators are convinced that their alarm systems are malfunctioning during an emergency when the alarms suddenly activate.

Monotony results from inactivity, boredom, or repetitive tasks. Monotony is difficult to measure, but its potential as a psychological stressor is readily apparent. The mind-dulling effects of monotony can seriously degrade performance.

Uneventful vigilance tasks are a prime example of tasks that produce stress via monotony. Closely aligned with monotony is sensory deprivation. Sensory deprivation occurs when a person is isolated from any perceptual input. Experimental studies of sensory deprivation reveal that perceptual isolation results in poor sensory-motor acuity and even hallucinations. The experience of monotony and sensory deprivation is no doubt stressful. The decreased alertness and poor sensory abilities that develop from monotony and sensory deprivation obviously contribute to degradations in performance.

Excessive task speed and task load requirements can cause psychological stress as well as physical stress. This stress is associated with the concern that the task be completed successfully within the time allotted. If successful completion of the task is required to ensure the safety of others, the stress is likely to be very great. Threats of failure and threats of job loss may compound this stress.

Disruptions in circadian rhythm can serve as both psychological and physiological stressors. The 24-hour cycle of night and day, or circadian cycle, is the basis for our regulation of eating, sleeping, working, and socializing. Adjustment to a circadian rhythm means adjustment of patterns of external temperature changes, solar radiation, and ambient noise levels. Disruptions of these cycles (e.g., changes in shift work) can result in autonomic nervous system changes and neuroendocrine changes that are stressful.

Such stress may impede psychomotor functioning and interfere with mental stability. Disruptions in circadian rhythms may also lessen endurance and produce irritability. Though a person may adjust to the pattern of a new shift within one week, it may be likely that his or her shift will be changed again very soon, causing another disruption.

A certain amount of fatigue and stress is inherent in every working situation, and it does not always affect work performance in a negative fashion. As mentioned earlier, a limited amount of stress serves to act as a motivator. However, when fatigue and stress are experienced in extreme states, their effect on worker output is deleterious. The quality of work may deteriorate, and the speed of performance may decrease. Frequencies of errors and accidents may increase as a function of stress and fatigue. The variability of a worker's performance may also increase. These changes in a worker's performance may have serious consequences in the execution of critical tasks, such as landing an aircraft, averting nuclear accidents, or properly assembling parts in a manufacturing process. It is the task of the human factors engineer to identify the situations that lead to extreme states of fatigue and stress, and to recommend how to avoid these states. Because of his or her knowledge of both humans and machines, the human factors engineer is well prepared to do this.

Finally, stress may be the result of the worker's life situation external to the work environment. Ergonomics studies indicate that management should be aware that such stress may also have a negative impact on work performance, and the worker may require some supportive attention for a period of time.

5 Seated and Standing Operations

M OST WORKERS SPEND almost all of their time in a small work area. It is important that this work area, or "work envelope," be well designed. The worker's physical relationship with this space has direct impact on productivity and physical and emotional well-being. Work tasks are performed while either sitting or standing, or possibly a combination of both.

SEATED WORK

Well-designed workstations for seated tasks have several things in common:

- Everything the worker needs to perform the task is available and easy to handle in the seated work area.
- Seated work does not require large forces (e.g., handling items that weigh more than 10 lb or 4.5 kg), although occasionally, mechanical assistance makes it possible to work with these weights while seated.
- A good chair is supplied, and nothing needs to be lifted from the floor.

The chair that a worker uses while performing the job should be chosen in relation to the job performed, and from the perspective of the worker. The chair should allow for the full range of motion required for the job task, allowing the worker to move around in the seat, lean forward, and get up and sit down easily.

Back stress is a major problem in seated work, and the nature of a chair's backrest is central to alleviating this difficulty. The backrest should be adjustable both horizontally and vertically, and should be set at an angle (usually 95 to 100 degrees). The height of the seat should be vertically adjustable. For some tasks, the chair might need forearm supports.

Here are some things to consider when evaluating chairs:

- Is the backrest adjustable vertically and horizontally?
- When adjusted to the worker, does the backrest clear the top of

the hip bone?

- How many legs does the chair have? (Five legs are more stable than four.)
- Does the chair have casters? Are casters appropriate for this job situation? (In some situations, casters may be a hazard.)
- Can a footrest be provided?
- Are armrests provided? Are they necessary?
- Are the adjustment mechanisms easy to reach and operate?
- Is the chair height adjustable?
- Is the seat height appropriate for the work space? (Does the chair allow the worker to sit comfortably with knees under the work surface?)
- Is there a front scroll edge on the seat?
- Can the chair be adjusted so the crease at the back of the knee is at or up to two inches above the height of the seat of the chair?

The workstation in which seated work is performed should be designed so everything in the work area can be grasped by both shorter and taller employees with a limited amount of forward leaning. Because distance from the body affects the ability to grasp and lift, the ideal work area should form a semicircle around the worker.

Seated tasks often involve lifting objects. Care must be taken with placement of the objects to be lifted. Even a relatively lightweight object lifted at arm's length can result in damaging forces to the lower back.

A seated worker resting comfortably in a chair with a good back support will have almost twice the hand strength with the hand turned inward than turned outward. When a worker is performing a task requiring rotational force (e.g., turning a hand wheel), the greatest force is achieved when the hand is grasping about 12 in. (30.5 cm) in front of the body.

The hand is more powerful when pulling downward than pulling upward, and has more strength when pushing than when pulling. Pushing power is greatest when the hand is about 20 in. (50.8 cm) in front of the body. Pulling works best when the object is grasped about 28 in. (71.1 cm) in front of the body.

A seated worker's arms should hang naturally at his or her sides in a relaxed position. The work surface should be at elbow height or slightly below it, so the forearm is horizontal or slightly slanted down.

The optimum work surface height for writing or light assembly work should be in the range of 27.5 to 31 in. (69.9 to 78.7 cm) above the floor. For heavier manual work, such as packaging, the work height should be 26 to 28.5 in. (66 to 72.4 cm) above the floor. The best height will vary from worker to worker, so it should be possible to adjust the work height.

STANDING WORK

Standing work is common in many industries. A great many jobs are performed with less exertion when the worker is standing. However, prolonged standing in one position creates different stresses.

Some of the problems of standing work can be alleviated by considering the following:

- If the worker stands for long periods of time on a hard surface like metal grating or concrete, fatigue-reducing mats may be provided. Some special types of cushion-soled shoes will also help.

- A general rule for work height (when the hands are performing the work) is 2 to 6 in. (5.0 to 15.2 cm) below elbow height. Elbow height is defined as the lower arm bent 90 degrees to the upper arm with the elbow resting along the sides of the body. This will alleviate shoulder and neck problems from a work height that is too high, and prevent upper body bending and back pain if the work height is too low. If the work height is not adjustable, in principle, design to accommodate tall employees and give shorter employees something to stand on as long as that does not become a safety hazard.

- The optimal height of the work surface may vary according to the task being performed. The best work height for light assembly, writing, or detailed inspection of parts would be higher than that for tasks requiring large downward or lateral forces. When heavy downward force is used, the work height would be lowest. A lower work height is also useful for tasks requiring large upward forces.

- Horizontal reaches to grasp tools, stock, and other materials should be prioritized by frequency of use. Frequently reached items should be located in a semicircular area extending about 14 to 16 in. (35.6 to 40.6 cm) in front of the body. Locate less frequently reached materials farther away from the body, but generally not more than 24 to 26 in. (61 to 66 cm) away from the body.

- Elevating one foot while standing can help alleviate low back stress. A bar, rail, or specially provided footrest can be used to alternate resting of the feet.

- The job should be structured to allow the worker to move his or her head frequently. The visual field and the operation should be limited enough to prevent excessive head movement, but not so limited that movement is impossible.

- When a standing worker is doing delicate work, the elbow should be supported to reduce back strain. A good work height for this type of work is 2 to 4 in. (5.0 to 10.2 cm) above elbow height.

COMBINED SEATED AND STANDING WORK

Some jobs require both sitting and standing, either because there is the occasional need to retrieve something from a location that is out of reach, or because several different tasks are performed, some of which are best done seated and others best done standing.

Workers who can alternate seated and standing positions are fortunate. They experience less fatigue from unvarying posture, and workstations designed to accommodate both will greatly improve the work situation. If possible, such workstations should be made with an adjustable work surface height. Another option is the use of prop stools. They allow for semi-standing and seated positions.

Another consideration in any work task is the dynamic or static nature of the job. In general, a task that allows greater movement overall will generate less physical stress than a job that requires little or no movement. Jobs that require bending to perform a task, or holding items for long periods of time are examples of the latter.

A static work situation is an important element in job difficulty whether the worker is seated or standing. Dynamic muscle activity enables nutrients to move into the muscle cells and waste products to move out of the cells. With static muscle activity, the muscles are contracted and the movement of nutrients and waste products is limited. As waste products accumulate in the muscle cells, a condition we call fatigue develops. An adjustment in the task that permits increased mobility is useful.

6 Manual Materials Handling

LIFTING OCCURS in a great variety of jobs, from jobs that include occasional light lifting to jobs that involve a great deal of lifting. Lifting is also a major cause of injury on the job.

CAUSES OF BACK INJURIES

What causes back pain? The following are the most common causes, in order of most frequent to least frequent:

1. Strain from improper sitting or standing postures, sometimes due to poor workplace design
2. Sudden or frequent twisting and/or bending of the back
3. Sudden strain on generally unused muscles, or jerking of the object
4. Lack of exercise, causing muscles to lose their strength, flexibility, and length
5. Manual materials handling tasks, such as lifting, lowering, pulling, pushing, and carrying
6. Stress
7. A "slipped" or herniated disc whose core has pushed against the nerves
8. Smoking (an aggravating factor)
9. Many different types of illnesses.

Structure of the Back

The spine is made up of bony structures called vertebrae held in place by ligaments (Figure 6–1). Because it is flexible, it allows for a variety of movements to take place. The functions of the spine are to support the head, protect the spinal cord, support some of the internal organs, and form an attachment for the ribs.

The spinal cord runs down the center of the vertebrae. Spinal nerves from the cord go to all parts of the body. As a result of poor use of the back, these nerves can be pinched, irritated, or damaged. Intervertebral discs made of cartilage act as cushions or "shock absorbers" between the vertebrae. As discs age or are subjected to wear and tear, they can suffer microfractures that cause them to lose their ability to cushion the spine. Discs impinging on nerves are an important source of body pain. The back muscles are designed for support and not for lifting. These muscles can be fatigued and stressed by improper use or overuse.

Pointers on Posture

The spine has three major curves: cervical—neck; thoracic—back; and lumbar—lower back (where most fatigue, aches, and pains occur). For correct posture, the position of the pelvis should be tilted slightly forward. This can be checked by standing against a wall with heels touching and the back flat against the wall. If the pelvis is not tilted forward, the lower back curve is increased.

If you stand with knees locked, the lower back curve is also increased. To stand for long periods of time, one leg should be up on a stool or rail. With the hips and knees bent or flexed, the lower back curve is decreased. To demonstrate this, lie on your back with knees bent. Put one hand under the lower back area and straighten the knees and hips. You should feel an increase in the curve.

Sitting posture. Use a proper chair, one that has adequate lumbar (lower back) support. The knees should be higher than the hips to

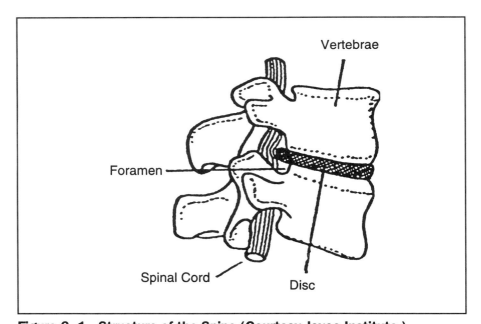

Figure 6–1. Structure of the Spine (Courtesy Joyce Institute.)

relieve lower back strain. A low box or foot stool can be placed under the feet. A good chair permits frequent body movement or flexing.

Driving posture. Position the car seat close to the pedals (reaching with the arms or legs puts a strain on the back). Also, with the seat close, more hip and knee flexion will occur. Keep the knees higher than the hips. Use the door armrest whenever possible. Avoid long periods of driving without periodic rest. If there is inadequate lumbar support to the lower back during long drives, roll up a towel and place it between the lower back and seat.

FACTORS AFFECTING LIFTING

Wherever lifting occurs, the focus of an ergonomics approach should be to design (or redesign) the task to make the most convenient way to perform the lift also the safest way. The following factors influence the ease and safety with which lifts can be performed:

- Force and amount of weight lifted
- Location of the load in relation to the body
- Size of the load
- Frequency of lifting
- Stability of the load
- Handles or handholds
- Geometry of the workplace
- Environmental factors
- Personal factors.

We will consider these factors one at a time.

Force and Amount of Weight Lifted

The best way to eliminate manual lifting injuries is to eliminate the need to perform the lift. If it is not possible to eliminate the lift, the next step is to keep the weight of the load at a safe level.

The definition of a safe weight level will vary from task to task and from worker to worker. The National Institute for Occupational Safety and Health (NIOSH) has developed an algebraic equation for determining acceptable and unacceptable lifting tasks. The equation takes several variables into account. A summary of these variables and how to measure them appears later in this chapter.

Location of the Load

Manual lifts can be performed most safely when they are closest to the body. The farther the load is from the body, the more pressure it exerts on the lower back.

The job should be structured in such a way that lifts do not have to be performed with one hand, or to the side of the body. The worker must be able to perform the lift without twisting the torso.

The vertical range within which lifts can be most efficiently performed is roughly 30 in. (76.2 cm) or knuckle height to the shoulders. If the task requires lifts from a lower level or to a higher level, everything possible should be done to redesign the workplace or to lift the load mechanically.

Size of the Load

The size and bulk of an object is as important as its weight. It can be more difficult to manually lift an unwieldy 20-lb (9-kg) object than to lift a 40-lb (18.1 kg) object that is compact and has good handholds. If the physical size of a load cannot be limited, large, bulky objects should be handled by more than one worker, or with mechanical assistance.

Frequency of Lifting

Frequency of manual lifting is directly related to energy expenditure. The more lifts a worker must perform during a given time period, the more muscle fatigue will occur. If lifting is frequent, even lighter loads will create a great deal of stress. The body needs time to recuperate between lifts; the likelihood of injury increases with fatigue.

Stability of the Load

If the center of gravity of a load shifts, a lift that began smoothly can lead to an overexertion injury. If at all possible, items to be lifted should be packed or arranged to avoid shifting of weight. If this is not possible, an alternative to manual lifting should be found.

Handles or Handholds

The handle or handhold is the coupling between the object being moved and the worker who moves it. The task should be designed with the hands of the worker in mind. That is, if an object is going to be lifted, it should accommodate the worker who will lift it. The type of handle or handhold used has a direct effect on the amount of force necessary to perform a lift.

There are essentially three types of grips:

1. The hook grip—the fingers are flexed around the object and the thumb is not used.
2. The power grip—the object is clamped between the partly flexed fingers and the palm, with the thumb opposing the grip along the plane of the palm.
3. The pinch grip—the object is pinched between the thumb and finger(s).

Handles or handholds that permit a power grip to be used offer the greatest mechanical advantage. Power grips also require the least amount of muscle force to be used in handling an object. Remember that the size and shape of the handles or handholds have a direct effect on the amount of strength required to perform a task. Figure 6–2 illustrates how the strength requirements of a task are affected by the size and shape of the handles.

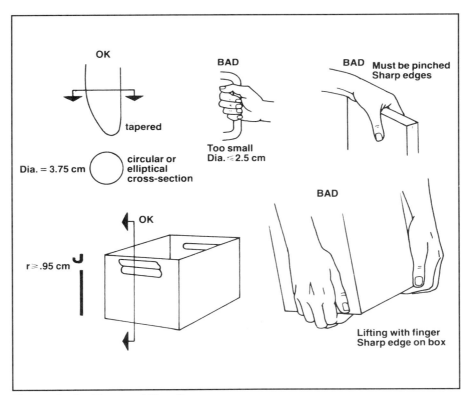

Figure 6–2. Types of Handles

Geometry of the Workplace

The design and layout of the workplace should take into account the need to perform lifts safely and smoothly. The shape of the work space will determine most of the factors involved in lifting (distance from the body, direction of the lift, postural constraints).

Workplaces should be designed with the following factors in mind:

- Lifts from the floor should be avoided.
- The torso should never twist while handling loads.
- Asymmetrical or unbalanced one-handed lifts should be avoided.
- Loads should not be lifted over obstacles.
- Loads should not be lifted at extended reaches.
- Loads should not have to be lifted with jerky or sudden movements.
- Uncomfortable postures should not be necessary throughout the work cycle. For example, it would be very poor design to arrange a task in which a worker would repeatedly lift objects off the floor onto a very low platform. This not only involves lifting off the floor, which is in itself risky, but it does not allow the worker to stand straight at any time during the work cycle.

Environmental Factors

Temperature, humidity, illumination, noise, vibration, and stability of the feet on the floor surface are all factors that contribute to the safety of a lift.

Workers expending physical effort in a hot environment perform less work and run the risk of overexertion and heat-related illness. Cold environments do not pose quite the same problem, but they may reduce manual dexterity and, if combined with vibration, may increase the risk of cumulative trauma disorders. In addition, humidity in very cold environments might be low enough to eliminate the moisture necessary to increase friction and grip effectiveness.

Illumination is important in all work situations. A worker who performs lifts needs enough light to see the work area and the object to be handled, and for adequate depth perception and contrast. It is also necessary for the worker to identify the weight load and center of gravity, either by seeing structural details of the object itself, or by reading the markings on it.

There should be frictional stability between the worker's foot and the floor surface. This can be achieved by making sure the floor is not too slippery or too rough, and that the worker is wearing appropriate shoes for the condition of the floor. The floor surface must also be clean and unobstructed.

Personal Factors

Both gender and age can affect a worker's ability to perform tasks involving manual handling of materials. The average woman's arm and torso lifting strength is about 60% of the average man's. Strength declines slowly with age, until age 65, when a person's strength is about 75% of what it was at its peak. Because the decline is gradual, it may go unnoticed. The people most at risk in lifting tasks are those between the ages of 30 and 50.

In spite of this decline in basic strength, the ability to work continuously does not diminish with age up to approximately 60 years of age in moderate environments. In order to utilize these capable workers on the job, a given task must be evaluated carefully for the worker's capabilities at the time.

TASK VARIABLES AND THE NIOSH LIFTING EQUATION

Each manual materials handling task involves certain task variables. These task variables describe a task and are used to determine if a manual materials handling problem exists.

The major task variables for lifting are:
- **Horizontal Distance (H)**—as the load mass center of gravity is moved horizontally away from the body, a proportional increase in the compressive force on the lower back is created. Thus, the further the load is from the body, the higher the stress.
- **Vertical Distance (V)**—the lower to the floor when the task commences, the more a worker must bend, thus placing a high stress on the lower back.
- **Vertical Travel Distance (D)**—the greater the vertical distance lifted, the more energy expended. Also keep in mind that people have less lifting strength above the shoulders.
- **Frequency of Task (F)**—the higher the task frequency, the more physical energy is expended. Frequency has a cumulative effect on the body. Beyond certain limits, frequency can result in fatigue, sprains, and strains.
- **Duration of Task**—refers to whether the lift is performed infre-

quently or continuously throughout the work shift. If the lift is performed infrequently during the work period, the body can recuperate its energy reserve, assuming the other job tasks are not physically demanding.

- **Weight of Load (W)**—weight is normally associated with lifting. The greater the weight, the more chance for lower back injury. The location of the weight handled is just as important as the amount of weight.

Each of these task variables is highly interactive. That is, the acceptability of load weight is dependent upon where the load is located (both horizontally and vertically), how far it is lifted, and how often. All of these variables must be considered before a lifting task can be said to be safe or unsafe under given circumstances.

In order to measure the variables, you will need a six- to eight-foot tape measure (1.8- to 2.4-m), a calibrated spring scale, and some sort of timepiece, such as a wristwatch or stopwatch. Here are the lifting task variables to measure and how to measure them (Figure 6–3). Measurements may be made in metric (meters and kilograms) or U.S. customary units (feet and pounds.)

- Horizontal Distance (H) is measured horizontally from the midpoint between the ankle bones to the midpoint of the handgrasp.

- Vertical Distance (V) is measured vertically from the floor to the midpart of the handgrasps at the beginning of the lift.

- Vertical Travel Distance (D), the distance of the vertical lift, is measured from the handgrasp at the beginning of the lift to the handgrasp at the destination of the lift.

When measuring these distances, do not have an employee hold a

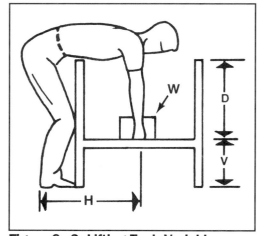

Figure 6–3. Lifting Task Variables

lift or maintain a hazardous position in order to take the measurement. This could cause an injury from overexertion.

- Weigh the object being lifted. If weights vary, record the average and maximum weights.

- Measure task frequency in terms of the number of lifts per minute.

- Figure the duration of lifting by determining if the lift is infrequent or occasional for one hour or less, or if the lifting is continuous over an eight-hour period.

All distances and weights can be rounded off to the nearest full unit. Accuracy of measurement is important, but rounding the variables will not compromise the results in terms of relative magnitudes.

THREE REGIONS OF LIFTING

The National Institute for Occupational Safety and Health (NIOSH) has defined three regions of lifting in *Work Practices Guide for Manual Lifting* (Figure 6–4). They are:

1. **Above the Maximum Permissible Limit (MPL).** These are considered unacceptable lifts requiring engineering controls. At this level only about 25% of healthy male industrial workers and fewer than 1% of healthy female industrial workers can perform lifts with reasonable safety at this level.

2. **Between the Action Limit (AL) and the Maximum Permissible Limit (MPL).** These are considered unacceptable lifts and require administrative or engineering controls. Administrative controls include such activities as employee training, employee placement, employee strength testing, and other adjunct programs.

3. **Below the Action Limit.** These lifting tasks present nominal risks to most industrial workers. At this level, more than 99% of healthy male industrial workers and about 75% of healthy female industrial workers can perform lifts with reasonable safety.

The purpose, then, is to determine at which level a particular lift falls. To do this, the following algebraic equation (developed by NIOSH) can be used. This lifting equation applies only to two-handed lifting in front of the body with no body twisting, to loads that are 30 inches (75 cm) or less in width, unrestricted lifting posture, with good coupling between the lifter, the floor, and the load, and to favorable ambient environments. (NIOSH is currently reviewing this formula and the *Work Practices Guide for Manual Lifting,* but the revision was not ready at the time of the printing of this manual.)

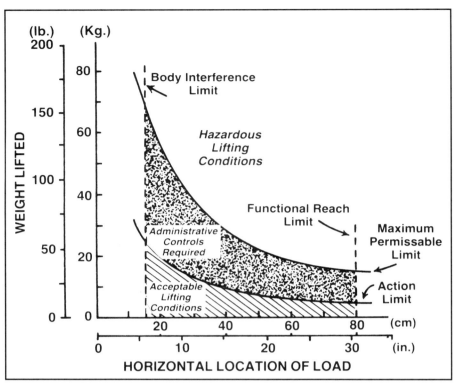

Figure 6–4. This illustrates the three regions of lifting defined for infrequent lifting (less frequent than one lift per five minutes) of a load 6 to 24 in. (15 to 60 cm) off the floor. Based on the horizontal hand location, the maximum weight that can be lifted (for the AL and MPL boundaries) can be determined. Note that the farther away from the body the load is, the less weight that can be lifted. (Courtesy NIOSH, *Work Practices Guide for Manual Lifting.*)

The formula in U.S. units (lb and in.) is:

Action Limit (lb) = $90(6/H)$ $(1 -.01 \mid V - 30 \mid)$ $(.7 + 3/D)$ $(1-F/F_{max})$

In metric units, the formula is:

Action Limit (kg) = $40(15/H)$ $(1-.004 \mid V-75 \mid)$ $(.7+7.5/D)$ $(1-F/F_{max})$

Maximum Permissible Limit (MPL) = 3 (AL)

> Where:
> | | | |
> |---|---|---|
> | H | = | Horizontal Distance |
> | V | = | Vertical Distance |
> | D | = | Vertical Travel Distance |
> | F | = | Frequency of Lift |
> | F_{max} | = | Maximum Frequency |

The maximum frequency represents the maximum lift frequency based on the vertical distance at the beginning of the lift. A worker bent over at the beginning of a lift (V≤30 in. or 76.2 cm) can lift less frequently than one standing and lifting (V>30 in.). From Table 6–1, select the proper F_{max} value, depending on the duration of the lift and the vertical distance at the beginning of the lift.

The following is an example of how to use the NIOSH lifting equation:

A worker is continuously lifting 50-lb (22.7 kg) parts out of a deep container for eight hours at a lifting frequency of once per minute. Is this an acceptable lift?

Measurements are taken for the task variables (Figure 6–5), which are:

20 in. = Horizontal distance   12 in. = Vertical distance
18 in. = Distance of lift   50 lb = Weight lifted
1 lift per minute = Lift frequency
8 hours continuous = Lifting duration
12 = Frequency maximum (from table)

The task variables are inserted into the equation and solved:

AL (lb) = 90 (6/H) (1–.01 | V–30 |) (.7+3/D) (1–F/F_{max})
AL (lb) = 90(6/20) (1–.01 | 12–30 |) (.7+3/18) (1–1/12)
          90 (.30)   (.82)   (.87)   (.92)
Action Limit (AL) = 18 lb
Maximum Permissible Limit (MPL) = 3(AL) = 54 lb

Table 6–1. F_{max} Table

F_{max} Table (Frequency Maximum)

	V ≤ 30 in.	V > 30 in.
		(Stooped) (Standing)
Infrequent	1 Hour	15	18
Continual	8 Hours	12	15

(Lift Duration)

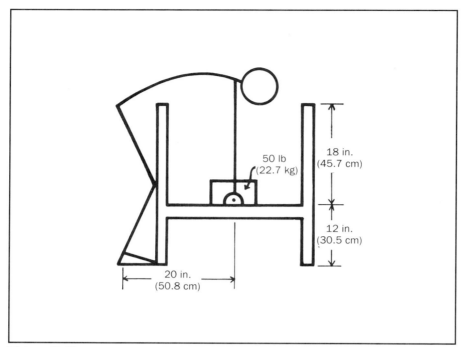

Figure 6–5.

The weight lifted (50 lb or 22.7 kg) falls between the Action Limit (AL) and Maximum Permissible Limit (MPL) established for that lift. According to the NIOSH lifting guide, this is an unacceptable lift requiring administrative and/or engineering controls. Because the 50 pounds lifted falls close to the MPL, the majority of males and females will not be able to perform this lift with reasonable safety.

Because different people have different tolerances to lifting, this lifting equation should not be used to determine if an individual can perform a certain lift. Instead, the focus should be on the lifting task. If a lifting task is above the AL, some of your workers will experience lower back injuries from performing the task, but you will not know which ones.

JOB DESIGN GUIDELINES FOR LIFTING TASKS

In general, when lifting occurs on the job, the work should be designed in such a way that the following elements apply:

- Workers assigned to lifting tasks should have training in lifting techniques and ergonomic principles.
- Lifts are smooth, that is, slow and without sudden actions.
- Lifting is performed with both hands and directly in front of the

body. There should be no twisting during the lift.

- The load should not be too wide (no more than 30 in. or 76.2 cm wide).

- The work environment should not restrict the posture of the worker performing the lift.

- The environment should be comfortable (e.g., not too hot or too noisy).

- Good couplings should be available, such as good contact between hands and handles and between shoes and floor surfaces.

- Workers who perform lifting tasks regularly should be required to perform only minimal amounts of other work, such as holding, carrying, pushing, pulling, or lowering.

- Workers who are assigned to lifting tasks should be physically fit and accustomed to physical labor.

The following is a broad range of ergonomic approaches that will help ease or eliminate problems in manual materials handling tasks. This list is not complete, but it is designed to stimulate ideas about how these and other ergonomic approaches can be applied to your workplace:

- Eliminate manual lifting and lowering tasks.

- Minimize the force required to move the load.

- Keep reach and lift distances as small as possible.

- Minimize the time.

Each approach is discussed below in more detail.

Eliminate Lifting and Lowering Tasks

The ideal approach is to eliminate manual materials handling through automation or mechanical assists. This can be done in several ways. Lift tables, lift trucks, elevating conveyors, hoists, and robots can all be used, as well as automatic feeding or air-rejection devices, gravity dumps, chutes, or conveyors (Figures 6–6 and 6–7).

If manual lifting cannot be avoided, then raise the load height so workers do not have to bend to lift. Set loads on platforms, tilt dollies, or specially built risers. Do not set loads on the floor that have to be lifted later. Not only may it be unsafe, it is also inefficient from a production standpoint.

If objects are just heavy enough to be lifted with some difficulty, discourage or prevent attempts to lift manually by increasing the weight to the point that the lift cannot be done manually and must be accomplished mechanically.

Minimize the Required Force

If the load is heavy or bulky, use mechanical devices. Conveyors may be used to transfer objects between surfaces without lifting.

Alter the task by using handles or hooks to secure a firm grip, using large wheels on carts to minimize the force needed to push or pull, balancing the contents of the container to stabilize the load, or treating the work surface to allow the load to move more easily.

It may be possible to construct containers of lighter materials, or to change the shape of the container to make handling closer to the body possible. If containers are used, split the load; distribute it in two or more containers for easier handling. If the load is still too heavy for one person to lift, design the job for two or more people.

Figure 6–6. A power lift table raises parts to a comfortable work height and reduces lifting distances.

Figure 6–7. A power parts dumper can raise materials from floor level and slide them onto a work table, eliminating bending over the container and lifting.

Try changing the nature of the job. Change lifting to lowering, change lowering to carrying, carrying to pulling, pulling to pushing. These changes allow for the assistance of hand trucks, dollies, conveyors, and so on.

Keep Reach and Lift Distances Small

The best way to keep distances small is to rearrange the task. Begin the lift as close to the body as possible. Heavy items should not be placed at any height lower than about mid-thigh or higher than the shoulders.

Store loads so they will be easy to retrieve. Locate loads within easy arm reach. This may mean lowering the work surface or raising the operator. Avoid deep shelves that would require a long reach and great pulling force to retrieve a load. Use spring-loading bottoms for bins or gravity bins to bring the load into easy reach.

Provide handles or grips, or change the shape or location of handles that are already provided. Try changing the shape of the container or object.

Minimize the Time

In addition, it is also helpful to reduce the amount of time involved in performing lifting tasks. This may mean changing the method that has been developed for performing the task, reducing the frequency of the lifts, or rotating jobs to less stressful tasks at regular intervals.

Rest breaks are vital for workers performing tasks that require strength and force. These should be a regular part of the work cycle to allow workers to maintain acceptable levels of physiological responses.

When the environment presents extremes in temperature or humidity, time should be allowed for the worker to adjust to the environment before performing heavy work. If extreme heat is present in the workplace, it is vital that heat stress be considered as a serious hazard. Be sure the job design takes this into account.

REDESIGN PUSHING, PULLING, AND CARRYING TASKS

The principles involved in improving pushing and pulling jobs are similar to those in lifting jobs:

- If possible, eliminate the need for pushing or pulling by using conveyors, slides, or chutes.

- Redesign or relocate the flow of work to keep the push, pull, or carry distance at a minimum.

- When pushing and pulling must occur, push whenever possible. Use large casters to make the job easier.

- Ramps should be limited to a slope no greater than 10 degrees.

- Handles and grips should be firm and comfortable, and located at a pushing height of between 35 and 50 in. (88.9 to 127 cm) off the floor.

- Be sure the load is low enough for good visibility when carrying loads, and that walking surfaces are properly maintained.

Whenever possible, replace carrying tasks with conveyors, hand trucks, carts, or dollies. If possible, use slides or tables between workstations to eliminate the need to carry or reduce distances. Rearrange the workplace to eliminate the need to carry.

When objects must be carried, reduce their weight as much as possible by reducing the size or capacity of the container, the weight of the container, or the load in it.

LIFTING TECHNIQUES

When lifting takes place on the job, it is vital that workers be trained on how and why to use good lifting techniques. The best workplace design will not protect a worker from injury if he or she does not understand and implement the principles of lifting when attempting a difficult lift.

No training program will guarantee that workers will perform their tasks properly. However, it can attempt to convince them to do so. The aims of training should be to make workers aware of the dangers of manual materials handling, to show them how to avoid unnecessary stress, and to teach them to be aware of what they themselves can handle safely.

Good training programs should involve basic principles of physics and functional anatomy, learning how to recognize what loads can be handled without undue effort and how they should be handled in relation to the body, and practice in acquiring the skills for safer and easier manual materials handling.

Lifting Recommendations

There are no comprehensive and sure-fire rules for "safe" lifting.

Manual materials handling is a very complex combination of moving body segments, changing joint angles, tightening muscles, and loading the spinal column. The following recommendations apply, however:

- Design manual lifting and lowering out of the task and workplace. If it nevertheless must be done by a worker, design the workstation so that the lifting can be done between knuckle and shoulder height.

- Ensure that workers assigned manual lifting tasks are in good physical shape. If the worker is not used to lifting and vigorous exercise, do not assign difficult lifting or lowering tasks.

- Place material conveniently within reach. Have handling aids available. Make sure sufficient space is cleared.

- Train workers to get a good grip on the load and test the weight before trying to move it. If it is too bulky or heavy, a mechanical lifting aid, an assistant, or both, should be used to help.

- Train workers to get the load close to the body, place the feet close to the load, stand in a stable position with the feet pointing in the direction of movement, and lift mostly by straightening the legs.

- Do not twist the back or bend sideways.

- Do not lift or lower awkwardly.

- Do not hesitate to get mechanical help or help from another person.

- Do not lift or lower with the arms extended.

- Do not continue heaving when the load is too heavy.

These recommendations can be applied to any lifting method, such as the two methods described in the next sections.

Two-Hand Squat Lift Method. This method can be used to lift small, compact loads that can be brought in between the knees and close to the body. This is not a good lifting technique if a load cannot be brought between the knees or if it involves repetitive lifting. Depending upon the lifting rate, repetitive lifting can result in fatigue.

Assisted One-Hand Lift Method. Manual lifting of objects out of large containers is a common practice in the manufacturing and warehousing industries. Many of these containers, such as box pallets and tubs, present lifting problems that could result in lower back injury because the two-hand squat lift method cannot be used.

For the most part, workers lifting parts out of containers tend to use two hands to lift. When the level of parts in the container gets low, the worker must stoop over to reach and lift the part. The lower the level of parts gets, the farther over the upper body must bend to reach the parts. This extremely bentover posture results in high levels of disc compression in the lower back region, and is particularly hazardous in repetitive lifting.

A good alternative to the two-hand squat lift is the assisted one-hand lift method. This method permits use of the large arm and shoulder muscles to perform the lift instead of the vulnerable muscles of the lower back.

In the assisted one-hand lift method, the worker rests one hand on top of the container, bends over to grab a part in the container, and then pushes down with the hand on the container top to force the upper body up to a vertical position. By using the non-lifting hand in the lift to raise the upper body, the stress is shared across the shoulders and arm, while reducing the stress to the lower back (Figure 6–8).

The assisted one-hand lift method is limited by the following:

- **Load weight.** Lifts should be kept to less than 30 lb (13.6 kg), and preferably should be less than 20 lb (9.1 kg).

- **Grasp area.** This method works only with objects that can be securely grasped with one hand.

- **Load length.** Loads longer than 18 to 20 in. (45.7 to 50.8 cm) tend

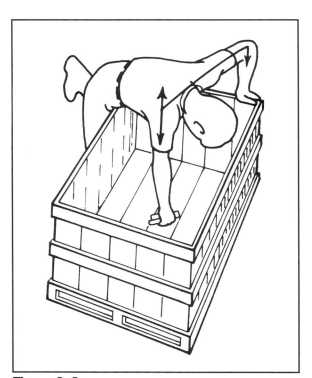

Figure 6–8.

to be awkward to grasp with one hand, and may stress the wrist if the lift is performed repeatedly.

The assisted one-hand lift is a viable alternative lift method to the two-hand squat lift when it can be performed. However, it should be recognized that the optimum solution to a manual lifting problem is through ergonomic design approaches. If the engineering approaches are not feasible, then emphasis can be placed on lifting methods.

The National Safety Council has information available on setting up complete training programs for manual materials handling, as well as detailed information on proper lifting procedures.

7 Occupationally Related Upper Extremity Disorders

Hand, wrist, and upper extremity injuries that are caused or aggravated by repetitive motion tasks and task stresses in industry are likely to be referred to as "cumulative trauma disorders" (CTDs). This chapter will discuss common causes of these disorders as we currently understand them and what can be done to reduce or prevent them.

COMMON RISK FACTORS

Risk factors associated with CTDs can be identified through careful analysis of a job. One worker may consider a task physically taxing, while another worker has no trouble at all. There are many factors that can affect a worker's vulnerability to CTDs. That is why there are no hard and fast rules for exposure limits.

The following generic risk factors are commonly associated with cumulative trauma disorders:

- High level of repetitive movement (less than 30 seconds per cycle, or more than 1,000 parts per shift)
- High amount of hand force (greater than 7 lb (3.2 kg) of hand force)
- Working with awkward postures (raised elbows and arms, bent wrists or hands)
- Mechanical stress (e.g., hands or forearms resting on a sharp table edge)
- High degree of vibration (e.g., from a power hand tool)
- Exposure to extreme temperatures
- A job that requires gloves
- Use of hand tools on the job (potential for working with awkward postures)
- A job that requires hand grasping or pinch gripping
- Certain types of grips (e.g., a pinch grip might require three or four times more force than a power hand grip).

Table 7–1 lists selected jobs and selected occupational risk factors. Not all jobs can be listed, nor will all jobs listed necessarily have these risk factors.

Table 7–1. Selected Jobs, Disorders, and Selected Occupational Risk Factors

Job	Disorder	Occupational Risk Factors
1. Buffing/Grinding	Tenosynovitis Thoracic outlet Carpal tunnel De Quervain's Pronator teres	Repetitive wrist motions, prolonged flexed shoulders, vibration, forceful ulnar deviation, repetitive forearm pronation
2. Punch press	Tendinitis of wrist and shoulder De Quervain's	Repetitive forceful wrist extension/flexion, repetitive shoulder abduction/flexion, forearm supination Repetitive ulnar deviation in pushing controls
3. Overhead assembly (welding, painting, auto repair)	Thoracic outlet Shoulder tendinitis	Sustained hyperextension of arms Hands above shoulders
4. Belt conveyor assembly	Tendinitis of shoulder and wrist Carpal tunnel Thoracic outlet	Arms extended, abducted, or flexed more than 60 degrees; repetitive, forceful wrist motions
5. Typing, keypunch, cashier	Tension neck Thoracic outlet Carpal tunnel	Static, restricted posture; arms abducted/flexed; high-speed finger movement; palmar base pressure; ulnar deviation
6. Sewing and cutting	Thoracic outlet De Quervain's Carpal tunnel	Repetitive shoulder flexion, repetitive ulnar deviation Repetitive wrist flexion/extension, palmar base pressure
7. Small parts assembly (wiring, bandage wrap)	Tension neck Thoracic outlet Wrist tendinitis Epicondylitis	Prolonged restricted posture, forceful ulnar deviation and thumb pressure; repetitive wrist motion; forceful wrist extension and pronation
8. Letter carriers	Shoulder problems Thoracic outlet	Carrying heavy load with shoulder strap

Table 7–1. *continued*

Job	Disorder	Occupational Risk Factors
9. Musicians	Wrist tendinitis Carpal tunnel Epicondylitis Thoracic outlet	Repetitive forceful wrist motions, palmar base pressure, prolonged shoulder abduction/flexion, forceful wrist extension with forearm pronation
10. Bench work (glass cutters, phone operators)	Ulnar nerve entrapment	Sustained elbow flexion with pressure on ulnar groove
11. Operating room personnel	Thoracic outlet Carpal tunnel De Quervain's	Prolonged shoulder flexion, repetitive wrist flexion, ulnar deviation (holding retractors)
12. Packing	Tendinitis of shoulder and wrist Tension neck Carpal tunnel De Quervain's	Prolonged load on shoulders, repetitive wrist motions, overexertion, forceful ulnar deviation
13. Truck driver	Thoracic outlet	Prolonged shoulder abduction and flexion
14. Core maker	Tendinitis of the wrist	Repetitive wrist motions
15. Housekeepers, cooks	De Quervain's Carpal tunnel	Scrubbing, washing, rapid wrist rotational movements
16. Carpenters, bricklayers	Carpal tunnel Guyon tunnel	Hammering, pressure on palmar base
17. Stockroom, shipping	Thoracic outlet Shoulder tendinitis	Reaching overhead Prolonged load on shoulder in unnatural position
18. Materials handling	Thoracic outlet Shoulder tendinitis	Carrying load on shoulders
19. Lumber/Construction	Shoulder tendinitis Epicondylitis	Repetitive throwing of heavy load
20. Butcher/Meat packer	De Quervain's Carpal tunnel	Ulnar deviation, flexed wrist with exertion

(From Vern Putz-Anderson, ed., *Cumulative Trauma Disorders: A Manual for Musculoskeletal Diseases of the Upper Limbs,* 1988: Taylor & Francis Inc., p. 22. Reprinted with permission.)

COMMON TYPES OF CUMULATIVE TRAUMA DISORDERS

Tenosynovitis

In this common type of disorder, the tendon sheaths in the wrists and fingers become sore and inflamed because of repetitive motions and awkward wrist position. It can be caused by poor workstation design, problems with tool design, or changes in work habits. The areas affected with tenosynovitis are often characterized by pain, swelling, cracking sounds, tenderness, and usually some loss of function. Tenosynovitis can be corrected by changes in work habits, tool design, or workplace layout.

Tendinitis

This condition is similar to tenosynovitis in both cause and symptoms, except that tendinitis is an inflammation of the tendon itself, with localized pain occurring at the site of the affected tendon or tendon group. The wrist, shoulder, and elbow are common locations for this injury.

Thoracic Outlet Syndrome, Bursitis, and Rotator Cuff Syndrome

These are disorders of the shoulder area that affect the nerves and muscles of the upper arms and shoulders. They are usually caused or aggravated by performing overhead tasks for extended periods of time.

Ganglion Cysts

This condition has often been associated with cumulative trauma or repetitive motion exposures. Ganglion cysts are tendon sheath or joint capsule disorders that often appear as bumps on the wrist. These bumps contain synovial fluid, and can be surgically removed.

De Quervain's Stenosing Tendinitis

This is an inflammation of the sheaths of certain tendons to the thumb. Scarring of the sheath restricts motion of the thumb. This condition usually occurs in people over the age of 30, and its cause is unknown.

Trigger-Finger Syndrome

This is another form of tendinitis, caused by repetitive flexing of the fingers against vibrating resistance. Eventually the tendons in the fingers become inflamed, and dexterity in the affected fingers may be reduced. Involved areas are often characterized by pain and swelling.

This condition may result from using a hand tool with a handle so large that the fingertip bends to supply force before the finger base, which is stronger, has an opportunity to help. When the condition is present, the flexor muscles are able to flex the finger against the mechanism, but the extensor may be too weak to straighten it out after locking. Manual extension of the fingers is usually accompanied by a clicking sound.

Hypothenar Hammer Syndrome

This problem is rare, but it can resemble carpal tunnel syndrome. It may occur in people who frequently strike objects repeatedly with the heel of their hands, as well as those exposed to long periods of extreme vibration. This can injure arteries above the carpal bones, restricting blood flow to the hand. The symptoms can be numbness of the fingers, insensitivity to cold, and a painful lump on the fleshy surface of the palm below the thumb.

Hand-Arm Vibration Syndrome

The use of electric or pneumatic tools to reduce the effort of manual tasks can result in exposure to vibration in the hands and fingers. Vibration sources in the 25–150 hertz range can result in Hand-Arm Vibration Syndrome (HAVS), formerly known as vibration white finger syndrome. Symptoms are numbness, pain and blanching of the fingers, loss of muscular strength in finger control, and increased sensitivity to heat and cold. Typical sources of this problem are pneumatic screwdrivers, hammers, chain saws, rotary grinders, and sanding machines. The adverse effects can be increased with increases in vibration, age, and length of exposure to cold environments.

Epicondylitis

Also known as tennis elbow, epicondylitis is an inflammation of tissues on the thumb side of the elbow. In an industrial environment, this disorder may be caused by tasks in which the forearm rotates the palm up against resistance, or by a violent or highly repetitive action, such as the use of a screwdriver. It may also be caused by extending the forearm horizontally and rotating the wrist and palm downward. This condition can be avoided by limiting the rotation of the tool or task to fit the natural limb rotation of the forearm. (A similar condition can occur on the little-finger side of the elbow.)

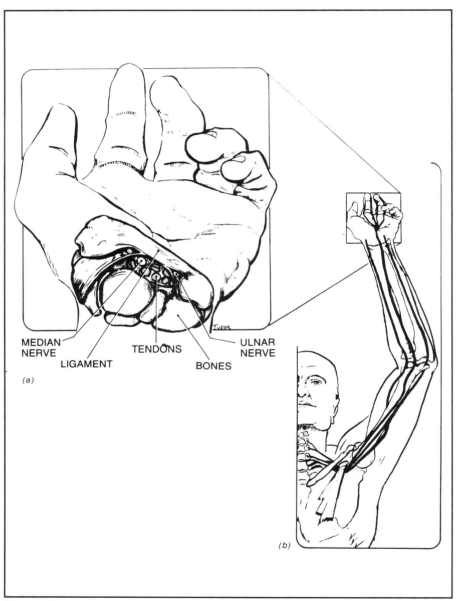

Figure 7–1. *(a)* A cross-section of the wrist showing the carpal tunnel that serves as a passage for finger tendons and the median nerve. *(b)* Pathway traced by three major nerves that originate in the neck and feed into the arm and hand: ulnar, median, and radial. (From Vern Putz-Anderson, ed., *Cumulative Trauma Disorders: A Manual for Musculoskeletal Diseases of the Upper Limbs,* 1988: Taylor & Francis Inc., p. 13. Reprinted with permission.)

Carpal Tunnel Syndrome

This condition is caused by repetitive overextension or twisting of the wrists, especially under force. It affects the median nerve, which runs through a channel called the carpal tunnel (Figure 7–1). If swelling should occur in this space, there is compression of the ligaments that surround the wrist, and the result is carpal tunnel syndrome (CTS).

Symptoms of CTS include burning, itching, and prickling or tingling feelings in the wrist or the first three fingers and thumb. There is usually some sensory change in these fingers. When the condition is severe, some muscles of the thumb may atrophy, and there is an overall weakness of the hand. Nerve conduction time may also decrease in the median nerve.

Occupationally related CTS occurs with equal frequency to men and woman of all age groups. Anyone who has symptoms suggestive of this condition should consult a physician familiar with the disorder. Treatment depends on the extent of the problem and other factors. Treatment can involve immobilizing the wrist with splints, providing support (combined with cessation of all activities that might be contributing to the syndrome), using anti-inflammatory medication, or, in severe cases, surgery.

Preventive Measures for Carpal Tunnel Syndrome

Because CTS has become a frequent problem in industry, this section will describe some ideas for prevention of this disorder (Figure 7–2).

Posture. Carpal tunnel syndrome and other cumulative trauma disorders can be minimized by use of proper hand posture, as well as overall good posture. Incorrect posture overloads the body's muscular structure. Job rotation can help to relieve stress in many cases, but the

---

Steps to Prevent Carpal Tunnel Syndrome

Overall Program	Ergonomic Aspects
• Placement	• Reduce frequency
• Training	• Reduce force
• Administrative	• Reduce deviation
• Engineering	• Reduce duration
	• Reduce restriction

Figure 7–2.

job to which workers are rotated should not use the same muscle group or cause other trauma.

Posture is important, not only for the hand and wrist, but also for the body, arms, and forearms. While performing a work task, the upper arms should be held close to the body, and the forearms should be angled slightly downward. Figure 7–3 shows how placement of

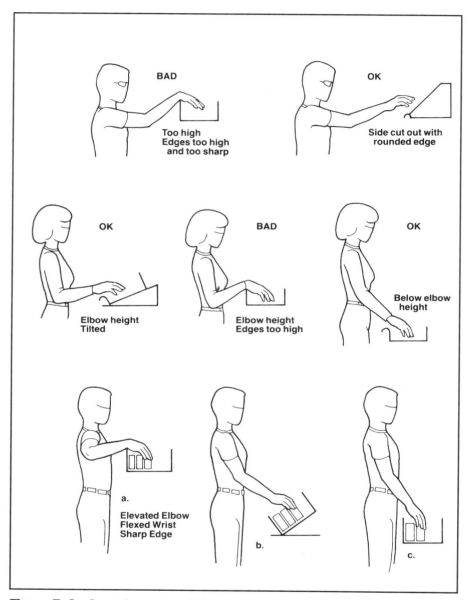

Figure 7–3. Containers should be designed so that workers can reach all locations without flexing their wrists. All edges that come in contact with workers should be well-rounded. A stressful work posture (a) can be controlled by relocating (b) or reorienting (c) the container.

containers can result in proper and improper posture of the arms.

The posture of the wrist should be in line with the hand (unbent natural position) to prevent pinching in the carpal tunnel. Figure 7–4 shows placement of jigs which result in proper and improper wrist alignment.

Grip. The way an object is gripped has a direct impact on the development or prevention of carpal tunnel syndrome. Gripping with the full hand rather than just the fingers allows the worker to use more power, reducing stress on the finger flexor tendons. Gripping pressure should be equalized over the entire palm of the hand as much as possible. This way, force will be spread over the hand. Grips with textured surfaces are recommended because they increase sensory feedback, allowing the worker to evaluate the force needed to perform a task more accurately. Gloves that are too thick or that fit poorly may reduce sensory feedback. Figure 7–5 demonstrates different types of grips.

Stress. The amount of stress required to perform a task should be minimized by changing the task or using different equipment. An example of undue stress is the screwdriver pictured in Figure 7–6, which puts a great deal of stress on the base of the palm because the handle design is inappropriate. If the screw must be installed mechanically, a larger handle will allow the stress to be evenly distributed over the hand.

Repetitive motion. Repetitive movement tasks are major contributors to carpal tunnel syndrome. Repetition increases the stress in any activity, and may cause inflammation or damage to the tendons. The only way to reduce repetition is to analyze the job task and devise a task design that can eliminate or minimize task repetition.

Vibration. Vibration is another important contributor. Power drills, saws, sanders, and buffers are all factors in carpal tunnel syndrome. Low-frequency vibrations (10–40 hertz) are particularly harmful.

The vibration hazard can be reduced by using the tools less frequently, isolating the hand from the vibration, or using tools that operate with a minimum amount of vibration, such as pneumatic or pulse tools. Some tools are available with devices that substantially reduce vibration.

Temperature. Temperature, particularly cold, is another risk factor. When the air is cold or the object being handled is cold, the hand should be isolated from the cold air or objects. (Heat has not been recognized as a significant risk factor for carpal tunnel syndrome.)

Force. When the amount of force required to do a task is not appropriate for the strength of the worker, the posture required, or the tools used, force becomes a factor in cumulative trauma disorders. The amount of friction between the hand and the object, and whether or not the worker is wearing gloves are also factors.

The hand is much stronger in some postures than in others. For instance, a power grip requires much less force than a pinch grip. The hand is strongest when the fingers are wrapped around the center of gravity of an object, rather than when the weight of the load is located at one end of the object.

It may be possible to reduce the amount of force required by altering the object being grasped. This can be accomplished by reducing the weight of the object itself, by picking up fewer objects at a time, by lifting the object with two hands instead of one, or by changing the size, shape, or the object itself.

Objects should be held in a balanced fashion, with hands grasping them at the center of gravity, so their weight does not tend to twist them out of the worker's hand. If the hand cannot be moved to the center of gravity, it may be possible to shift the center of gravity by reducing, shifting, or adding weight to one side of the object.

Friction combined with weight can increase the force required to

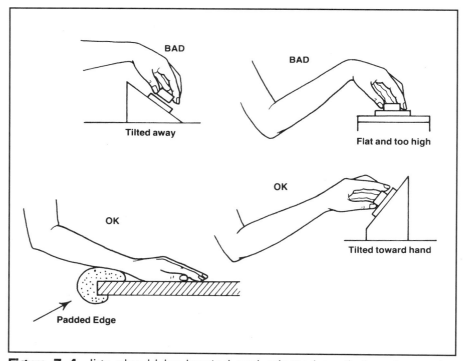

Figure 7–4. Jigs should be located and oriented so that parts can be assembled without flexing the wrist.

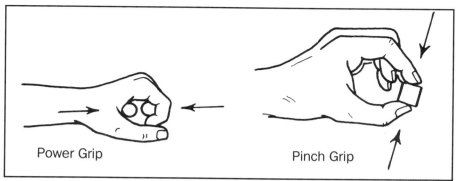

Figure 7–5. Different Types of Grips

move an object. It is possible to control friction by careful selection and texture of handle and glove materials. It is also helpful to keep handles free of grease and oil, and maintain the moistness of the skin.

Gloves reduce the amount of power available in the hand. It is helpful to get gloves that fit well, although it can be difficult to find gloves that fit really well because of the wide differences among people's hands. The best solution is to choose gloves that cover only the area needing protection. If the palm needs protection, cut off the glove fingers. If only the fingers need to be protected, they can be wrapped with safety tape, instead of wearing gloves.

Mechanical stresses. Mechanical stresses can produce tendon disorders as a result of the hand's contact with hard, sharp objects, or by pounding. For example, trigger-finger syndrome is often associated with using tools that have hard or sharp edges on their handles (Figure 7–8).

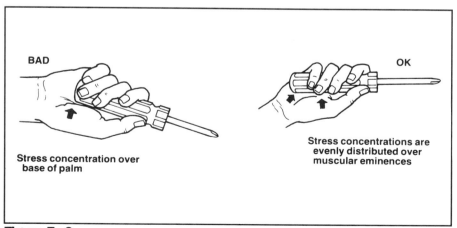

Figure 7–6.

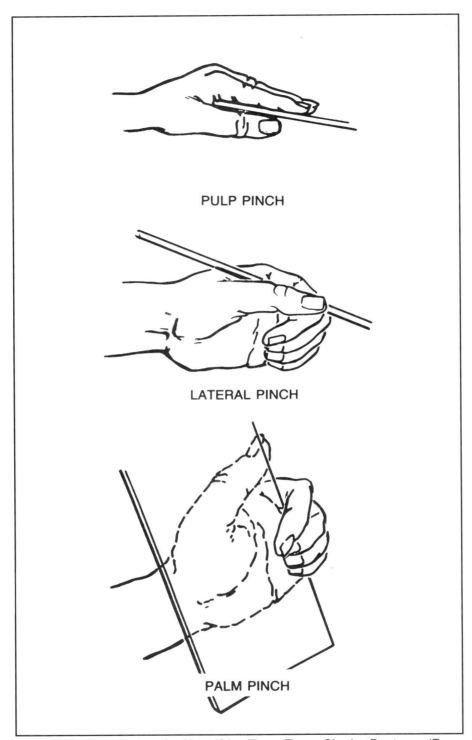

PULP PINCH

LATERAL PINCH

PALM PINCH

Figure 7–7. Terminology for Classifying Three Finger-Closing Postures (From Vern Putz-Anderson, ed., *Cumulative Trauma Disorders: A Manual for Musculoskeletal Diseases of the Upper Limbs,* 1988: Taylor & Francis Inc., p. 56. Reprinted with permission.)

It is possible to control the concentration of stress in the hand by increasing the size of handles, eliminating or rounding sharp edges, and using materials that "give." Handles should be as large as will fit comfortably in the hand, and this evaluation will vary according to the force required and dexterity involved in a particular task.

Any edge the hand touches should have as large a radius (rounded corner) as possible. This includes the edges of tables and jig fixtures with which the worker's hand or arm comes into continual contact. Padding should be added to ease this stress (Figure 7–9). These stress points can often be located by the presence of worn or polished surfaces where the worker's arm has continually rubbed against the surface.

Pounding with the hand should be eliminated, if possible. In some cases, the pounding can be done with a hammer or other percussive tool. If it is not possible to eliminate the pounding, padding should be used to cushion the stress.

BONE   TENDON   SHEATH   SKIN

Figure 7–8. A hand tool can compress finger tendons *(a)*. A sharp-edged tool *(b)* is more likely to produce injury than a round tool *(c)*. (From Vern Putz-Anderson, ed., *Cumulative Trauma Disorders: A Manual for Musculoskeletal Diseases of the Upper Limbs,* 1988: Taylor & Francis Inc., p. 66. Reprinted with permission.).

Job rotation. Job rotation should be considered to alleviate the stresses of jobs that cannot be modified by ergonomic intervention. The job to which workers are rotated should not use the same muscle groups or cause other trauma. Workers should be rotated to jobs that require different types of movements.

The key to preventing and controlling carpal tunnel syndrome disorders is early intervention—that is, early reporting and treatment. Supervisors and workers should be trained to recognize the symptoms of carpal tunnel syndrome. Initial symptoms include numbness, tin-

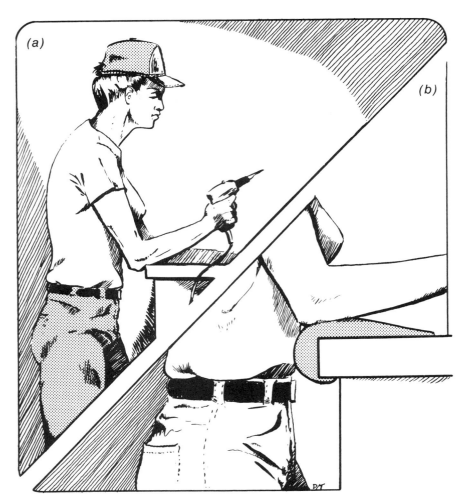

Figure 7–9. *(a)* A work posture that places pressure on the ulnar nerve. *(b)* A rounded pad placed over the edge of the work table relieves pressure on the ulnar nerve as it passes over the elbow. (From Vern Putz-Anderson, ed., *Cumulative Trauma Disorders: A Manual for Musculoskeletal Diseases of the Upper Limbs,* 1988: Taylor & Francis Inc., p. 96. Reprinted with permission.)

gling, lack of finger sensation, and muscular weaknesses.

Early treatment may include restriction of job activities and wrist splinting. With the advice of a medical practitioner, ice packs during the initial inflammatory stage can be followed by the use of heat packs. The worker should not be returned to the same task that produced the symptom.

8 Tools, Controls, and Displays

U SE OF POWER AND NONPOWER HANDTOOLS may involve risk factors that can result in cumulative trauma disorders. The risk factors associated with hand tool problems (not necessarily in order of concern) besides task repetition are hand and wrist positions, force required to manipulate the tool, grip and handle design, twisting (torque), vibration, pressure points, and static muscle loading. (See also Chapter 7 and Appendix 4.)

AWKWARD HAND AND WRIST POSITIONS

Grip strength is greatest when the wrist is straight (Figure 8–1). As the hand moves away from that posture, stress increases on the nerves and tendons entering the hand. When the wrist is bent and force must be applied, cumulative trauma disorders can result.

Awkward wrist positions can be caused either by the nature of the tool handle or the position of the object being worked on.

Several tool manufacturers offer tools that can improve the situation ergonomically in certain awkward tasks. These tools are usually bent to allow the tool to assume the awkward position, rather than the hand. However, these tools often have a limited range of use. Modification of the tools in the workplace is often necessary to achieve a specific result.

When evaluating and correcting tool-induced awkward hand positions, some combination of the following will be useful:

- Use of ergonomically designed tools
- Modification of the work task to properly use tools
- Modification of the work surface to angle the work object in order to increase its accessibility
- Reduction of the force necessary to operate hand tools, which includes switching to power tools, if possible.

The proper tool depends upon work orientation, work methods, and work heights. Hand tools should be supplied so that the user can grasp, hold, and manipulate the tool without bending the wrist.

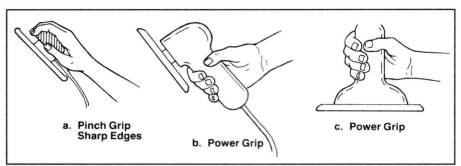

a. **Pinch Grip Sharp Edges**

b. **Power Grip**

c. **Power Grip**

Figure 8–1.

Straight Line vs. Pistol Grip

Figure 8–2 shows several correct uses of straight-line drive power tools and pistol grip tools in which the operator's wrist has been kept as straight as possible.

Straight-line tools that are used in a vertical downward force application should have a flange or sleeve at the bottom on which to rest the hand to help reduce grip forces.

FORCE REQUIREMENTS

The force required to perform a task is related to the posture of the hand, the characteristics of the object grasped, and the amount of friction between the hand and the object. See the section on Cumulative Trauma Disorder in Chapter 7 for a discussion of force as a factor in the disorder.

Force can be reduced in a variety of ways. The tool handle can be altered to make it more efficient, or the weight of the object being held can be reduced. The task can also be altered to require less exertion with each action (e.g., picking up fewer objects at a time, using a tool to work on material that is less dense).

Tools should be held at their center of gravity, so their weight does not twist them out of the hand. When the hand location cannot be changed easily, it may be possible to change the center of gravity by reducing the weight of the tool, shifting its weight, or adding weight to the light end.

Force requirements are greater for nonpower tools than for power tools. Take extra care to avoid awkward hand positions and pressure on tissues or joints. With nonpower tools it is also extremely important to be certain the handle fits the nature of the user's hand.

GRIP AND HANDLE DESIGN

As was discussed earlier, the way an object is gripped is important in preventing wrist problems. Good grip design for hand tools will keep the worker's wrist as straight as possible. In addition, tool handles should have a positive stop or flanged end to increase the stability of the hand. There should be no sharp corners or edges.

Try to avoid form-fitting handles, such as those with finger grooves, as the form they fit is probably not the form of the person using the tool.

Textured handled surfaces are useful, because they aid in sensory feedback. That allows the worker to assess the force needed to perform a task. Gloves may reduce specific stresses, but they also reduce sensory feedback, and may reduce the amount of strength available to

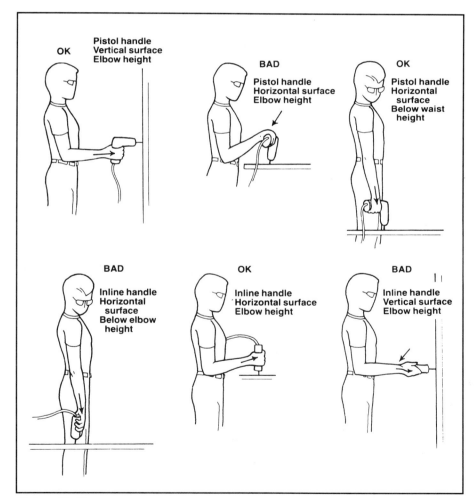

Figure 8-2.

apply to a task. If gloves must be used, they should cover only the area needing protection.

If possible, find an alternative to the use of a power tool that is started and stopped by a one-finger device. A one-finger actuating mechanism on power hand tools can cause trigger-finger syndrome, depending upon the force it takes to activate the tool and how often the tool is activated. This type of device can be replaced with a lever arm that is manipulated by several fingers, air-operated start methods, push-start methods on the drive shaft, or a switch that is operated by the thumb, rather than a finger. (See Trigger-Finger Syndrome in Chapter 7.)

A power tool that has extensions or additional long cords or air lines is unbalanced, and this unbalanced condition increases stress for the worker. Wherever possible, these tools should be balanced. A neutral-force tool balancer is better than one that automatically reels the tool upward. This eliminates the need for continual reaching and pulling the tool to the work area. Figure 8–3 illustrates the features of an air-powered, straight-line hand tool.

TORQUE

Driving a fastener into material with a power tool is likely to transfer torque to the worker's hand when the fastener bottoms out. This is experienced as a snapping action, and the repeated stress is a potentially serious problem.

This stress can be alleviated by using slip clutches or torque limiters, torque-reaction bars that keep the torque settings low, or by mounting the tool on an articulating arm to keep the torque from reaching the hand. Another solution is to provide an extra handle so the worker can use two hands to help counter the torque effect.

VIBRATION

Some workers who are continuously exposed to the vibration of power tools may experience circulatory problems. The length of exposure to the vibration of power tools is a major factor, as is the vibration frequency. If possible, frequencies between 40 and 90 hertz should be avoided, although low-frequency vibrations are also harmful.

The effects of vibration on hand tool users can be reduced in the following ways:

- Purchase low vibrating tools.
- Balance machines dynamically.
- Reduce speed.
- Limit exposure times.
- Ensure that machines are mounted stably and are properly maintained.
- Rotate personnel to reduce exposure time.
- Improve cushioning, suspension, vents, etc.
- Use properly fitted gloves to help damp the vibration.
- If possible, alter the task to reduce vibration.

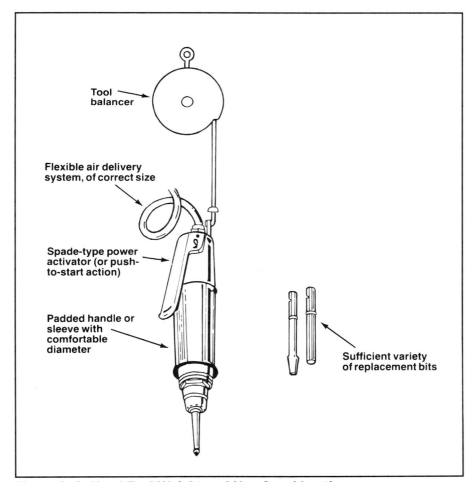

Figure 8–3. Hand Tool Weight and Use Considerations

NONPOWER HAND TOOLS

With the exception of vibration and switch actuation methods, the problems associated with nonpower hand tools are similar to those of power tools. These problems include awkward hand positions, stress on tissues or joints, excessive force requirements, and handles that are not appropriate for the user's hand.

Depending on the user's hand size, the handle grip span for such tools as pliers, wire strippers, or scissors should not exceed 2.5 to 3.5 in. (6.4 to 8.9 cm) (Figure 8–4). Tool handles should be long enough to extend past the palm of the user's hand. This length should be 5 in. (12.7 cm), with a minimum length of 4 in. (10.2 cm). If the worker is wearing gloves, another half inch of length is needed.

Hand tool handle diameters should be a nominal 1.5 in. (3.8 cm). Less force can be exerted with a narrow-handled tool than with a wider one. In addition, wider-handled tools distribute pressure over a wider area in the palms of the hands. If used repetitively, narrow-handled wrenches, for example, should have the handles built up to be held comfortably in the hand.

Tool handles should not have sharp corners or edges. Handles like those on pliers or scissors should be designed to open with a spring or other device. This avoids the exertion of force with the sides or back of the fingers. Avoid using hand tools with finger grooves because they seldom fit the user and could damage the nerves in the sides of the fingers if used often enough.

CONTROLS AND DISPLAYS*

Controls and displays that are well-designed and properly placed will help reduce accidents and stress injuries (Figure 8–5). The following general principles should be considered in selecting, designing, and placing controls and displays:

Guidelines for Controls

- The most frequently used controls should be within easy reach.
- All controls should be placed or guarded so that they will not be accidentally activated.
- The number of controls should be kept to a minimum.
- Assign controls that require precision during high speed operation to the hands. When there is only one major control that may be

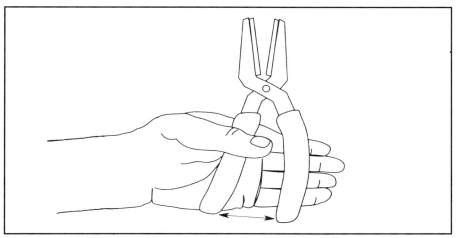

Figure 8–4. In this ergonomically designed tool, the grip span should not exceed 2.5 to 3.5 in. (6.4 to 8.9 cm).

operated by either hand or both hands, place the control in front of the operator—midway between the hands.

- Assign to the feet controls that require the application of large forces; otherwise, provide power-assisted hand controls.

- Distinguish between emergency controls and those that are required for normal operation through separation, color coding, labeling, or guarding. Emergency controls should be easily accessible and within 30 degrees of the operator's normal line of sight.

- To prevent accidental activation of a control, place it away from other frequently used controls, recess it, or surround it with a shield.

- Provide tactile (sensory) feedback to indicate that the control has been activated. Controls without this feedback (e.g., membrane switching) should have auxiliary audible or visual signals.

- Hand controls should operate on a pattern in keeping with normal human expectations (i.e., do not surprise operators).

- Consider size, shape, and color coding for proper control emphasis.

- Control height should be located between the top of the shoulders to the bottom of the elbows.

*Adapted from *Ergonomic Design for People at Work,* 1983 and Humanscale, 4/5/6, 1981.

Guidelines for Displays

Viewing distance. Many displays are designed for reading at arm's

---

Ergonomic Principles for Controls and Displays

Quick, error-free decisions and actions depend upon:

• Displays that are easy to read

• Controls that are easy to identify

• Controls that react the way we expect them to

• Feedback telling the worker how effective his or her decisions and control actions are.

Figure 8–5.

length to permit the operator to read switches and adjust knobs. This limit is generally set at 28 in. (71.1 cm) and is used to determine the recommended dimensions for display scale markings and readings.

Illumination. The characters on displays should be in high contrast against the background and should be located to avoid reflected light (glare).

Angle of view. The preferred angle is 90 degrees to the plane of the display (further than this can cause parallax distortion).

Presence of other displays. Displays should not look too much alike because the operator may read the wrong one. They should also be labeled clearly.

Compatibility with related controls. Displays and controls should be designed and located so that the untrained operator will select the correct control and operate it as expected.

9 How to Implement an Ergonomics Program

Wer AN ERGONOMICS PROGRAM is developed, it should be organized into a formal program created on an existing program structure, if possible. An organized program will get more results than a fragmented or disorganized one. This chapter is a summary of methods and procedures to set up an ergonomics program. See also Appendix 5 for a list of steps in developing an ergonomics program.

The first steps toward implementing a successful ergonomics program are top management commitment and employee involvement. Without these, success will be difficult. The next step is to identify problem work environments and processes. Through careful analysis of available data, you will be able to set priorities and define program needs, even without the help of a staff specialist on ergonomics.

Once the highest priority jobs have been identified, a plan of action should be developed to improve the work situation in those jobs. The plan should be easy to implement and should be oriented toward results and cost effectiveness.

TOP MANAGEMENT COMMITMENT

There must be a demonstrable commitment by top management to show that they support the ergonomics effort and provide the motivating force necessary to deal with the problem. They must establish company policy, assign responsibilities, and provide the resources to carry out the program. The program should be in writing and should provide for an annual program review and evaluation.

Before making a substantial commitment of time and money to an ergonomics program, top management will want solid evidence of the need for the program. Be sure to document any problems as clearly as possible. Offer suggestions on action to be taken to solve the problems.

If possible, equate ergonomic changes with return on investment (ROI). This can come about through gains in productivity, energy savings, product modification savings, and reduced costs. Some intangibles—improved employee morale, for example—can also be factors.

Plant managers are likely to be concerned about the reaction of the workforce to the news of plans for change. Be sure recommendations are intended to make the work process go smoothly, safely, and economically. The advantages and disadvantages of program implementation must be weighed carefully. Clearly thought out, positive plans will help calm employees' anxiety about changes.

EMPLOYEE INVOLVEMENT

Representatives of every area involved in the program should participate in designing and realizing the change. Such a group, which combines individual insights and perspectives, can bring far more knowledge and experience to the process than any individual working alone.

If a workplace is being redesigned, every element should be considered. That includes safety and health, equipment and tooling costs, and the technical capabilities of the system. An expert on ergonomic analysis, design, and testing/evaluation disorder and an engineer could work together for a design solution that takes into account an operator's needs, technical requirements, and cost considerations.

Here are some questions to help you decide whether or not a group approach will work for your situation:

- Is time a critical factor? (If it is, group participation will not work; groups act more slowly than individuals.)

- Is management committed to implementing the solution that the group recommends? (A group approach will not work unless the answer is "yes.")

- Is management already committed to a particular solution? (If they are not, participation can work.)

- Do workers have enough knowledge to make a good contribution to the process? (If they do, participation can work.)

- Do workers share management's goals? (If so, participation can work.)

- Is worker acceptance important? (If so, they should participate.)

- Do the workers want to participate? (If they want to, they should.)

Employees can provide excellent suggestions for task improvements or job modifications. No one knows a job better than the person who performs it day in and day out. Workers who have helped make a change tend to become committed to that change. They have a personal stake in seeing that the change is successful. Workers who have helped create the solution to a problem will become part of the solution, instead of obstacles to its success.

Most workers would like to have more influence over the environment in which they work. Ergonomic planning is an ideal opportunity for worker participation, because the workers will understand and identify with the goals of improving workplace safety and health. In addition, if the workers contribute to the new design, they are more likely to accept it.

SETTING UP THE ERGONOMICS PROGRAM

Everyone involved should be accountable for specific tasks and responsibilities to bring about ergonomic improvements in the workplace. If necessary, people should be trained in any new procedures they will be using after changes are made.

The people who are actively involved in creating the plan for ergonomic improvement should be formed into an ergonomics committee. The committee should be headed by a trained ergonomics coordinator.

This training should include principles of biomechanics, industrial engineering, physical anthropology, and other related subjects. The ergonomics coordinator should have access to a comprehensive library of ergonomics textbooks, scientific journals, and data bases.

The committee might also include employee representatives and representatives of plant safety, plant engineering, industrial engineering, medical, maintenance, industrial relations, and production departments.

The committee should evaluate the results of job task analysis, review the patterns and trends of cumulative trauma injury, and consider employee complaints. Following that, the committee should:

- Establish priorities for addressing problem areas
- Propose solutions to problems on various jobs and processes as determined by injury trends
- Analyze the potential effectiveness of proposed ergonomic solutions
- Encourage employee participation in the ergonomics program at all levels of the organization
- Review training plans for management, supervisors, professional staff, and hourly employees on the meaning of ergonomics and its purpose in the workplace

- Review research on special problems
- Develop and propose the strategy and goals of the ergonomics program.

In addition to functioning as a member of the committee, the ergonomics coordinator might:

- Identify cumulative trauma injury experience and set priorities for reducing or eliminating the problems
- Perform job studies to identify causal factors and develop solutions
- Monitor the progress of the ergonomics program and report the results to management
- Guide the committee in developing and proposing ergonomics strategy and goals
- Develop awareness and application training programs for engineers, special staff, managers, supervisors, and other employees
- Build an ergonomics reference library.

If a consultant is to be hired to help develop an ergonomics program or perform a job analysis, see Appendix 6 for guidelines for selecting the consultant.

The U.S. Department of Labor has issued *Ergonomic Program Management Guidelines for Meatpacking Plants*, which may serve as a quick guide in identifying key program elements.

PROBLEM IDENTIFICATION

Problem identification is an important part of the ergonomics program. If ergonomics problems are not specifically identified as to location and trends, and patterns are not identified, then any effort designed to reduce these problems will not be fully effective. "Shotgunning" solutions to imagined problem areas is an indication of an ineffective ergonomics program. Review Chapter 2 for a more detailed discussion of identifying problem areas.

HAZARD PREVENTION AND CONTROL

An important part of every ergonomics program is hazard prevention and control. After the job has been evaluated and the specific problems identified, the next step is to correct the problem. This can be done through a combination of engineering controls, work practice controls, and administrative controls.

The most positive thing that can be done to prevent the problem from occurring again is to redesign a job to eliminate the problem. The focus of an ergonomics program is to make the job fit the worker, not to make the worker fit the job. This is accomplished by redesigning the workstation, work methods, or tools to help the worker avoid applying excessive force and repetition, and assuming awkward posture.

Workstations should be designed to accommodate the vast majority of the workers for a given job. Work practice controls should address establishing provisions for safe work practices/procedures that are understood and followed by managers, supervisors, and workers.

Administrative controls should address the frequency and severity of exposures to ergonomic stressors. Some of the controls that might be used include: varying work routines, increasing the number of employees on the job, job rotation, and providing break-in/muscle conditioning time.

TRAINING

For any ergonomics program to succeed, the people involved must understand how to identify and control excessive physical stress in industrial operations. If a group approach is being used, all participants should get some training in ergonomics.

Training and education is a critical component of an ergonomics program. Through training, managers, supervisors, and workers come to understand the ergonomic and other hazards associated with a job or production process. A comprehensive training program addresses workers in the following categories: affected workers, supervisors, engineers, and managers.

The training program should deal in ergonomic problem identification and problem solving. It should describe, in general, the varieties of cumulative trauma disorders and lower back overexertion injuries, the means of prevention, its causes, and early symptoms.

Training should be periodically monitored to test its effectiveness through the use of worker interviews, oral testing, written testing, and observing work practices.

Appendix 1
Overall Facility Checklist

A. Indicators of the Need for Ergonomic Engineering Evaluation

☐ Is a new production line or facility being considered?

☐ Is production efficiency too low?

☐ Is product quality low?

☐ Are absenteeism and accident rates unusually high?

☐ Are back injuries or cumulative trauma disorders of the hand occurring frequently?

☐ Are medical visits occurring too frequently?

☐ Is turnover at the facility too high?

☐ Is turnover for specific tasks especially high?

☐ Does it take too long to train workers for certain tasks?

☐ Do workers make frequent mistakes?

☐ Is there too much waste material resulting from production?

☐ Is there too much equipment damage?

☐ Are workers frequently away from their workstations?

☐ Are employees making subtle workplace changes?

☐ Are workstations used during more than one shift each day?

☐ Are your plant engineers familiar with ergonomic principles?

☐ Do you utilize an incentive pay system?

☐ Do the employees seem to exercise their hands, fingers, or arms often to relieve muscle strain?

B. Indicators of the Need to Redesign Specific Tasks

☐ Are workers frequently required to lift and carry too much weight?

☐ Do workers have to push or pull objects that require large breakaway forces to get started (e.g., carts, boxes, rolls of material)? Do workers push or pull hand trucks or carts up or down inclines or ramps?

☐ Does a job require a worker to push, pull, lift, or lower objects while the body is bent, twisted, or stretched out?

☐ Do workers complain that they do not get enough breaks? Is the work pace not under the worker's control? Is this pace rapid?

☐ Does the task require the worker to repeat the same movement pattern at a high rate of speed?

☐ Does the worker's pulse rate exceed 120 beats per minute while doing the job?

☐ Is the job overly monotonous?

☐ Does the job involve the frequent use or manipulation of hand tools?

☐ Does the task require the continuous use of both hands and both feet in order to operate controls or manipulate the work object?

☐ Does the job require the worker to raise arms above shoulder height often or for extended periods of time?

☐ In order to perform the task, must the worker maintain the same posture (either sitting or standing) all the time?

☐ Does the job require the worker to keep track of a changing work situation mentally? Does this work situation require monitoring several machines?

☐ Must the operator process information at a rate that might exceed his or her capability?

☐ Must the operator sense and respond to information signals occurring simultaneously from different machines without sufficient time to do so?

C. Indicators of the Need to Redesign the Workplace

☐ Do workers sit on the front edge of their chairs, not using back supports?

☐ Do workers frequently add cushions and pads to their work chairs?

☐ Is it necessary for the worker to get into an unnatural or stretched position in order to see or reach gauges, controls, dials, materials, or part of the work object?

In order to operate foot pedals while standing:

☐ Does the operator have to operate foot pedals or knee switches?

☐ Must the worker assume an unnatural or uncomfortable posture?

☐ If there are foot pedals, are they too small to allow the operator to alter the position of the foot?

☐ Is a raised footrest necessary?

☐ In order to perform the task, must workers hold their arms or hands up without armrests?

☐ Is it difficult to operate controls or observe dials?

☐ Are dials or controls poorly labeled?

☐ Is the equipment designed or placed in such a way that cleaning and maintenance activities are difficult?

☐ Does there seem to be too much clutter in the workplace?

☐ Must the worker perform his or her other job in a chair that cannot be adjusted?

☐ Is it possible to provide clamps or supports that will relieve the worker of the need to hold the work object while performing the task?

D. Indicators That Special Considerations Need to Be Made in the Work Environment

☐ Is there so much process noise that hearing loss could occur?

☐ Is there so much noise that it interferes with speech or audible signals of various kinds?

☐ Is special lighting necessary to perform the job?

☐ Is there a sufficient difference between the background color for the task and color codes on knobs, handles, and displays?

☐ Does the job require the worker to look from dark to light areas on a regular basis?

☐ Are there sources of direct or reflected glare in the work area?

☐ Do lights reflect off machinery, causing distancing flashes or stroboscopic effects?

☐ Is the air temperature too cold? Too hot?

☐ Is it too humid in the workplace?

☐ Are radiant heat sources placed near any workstations?

☐ Are there rapid changes in temperature or light in the work environment?

☐ Is there sufficient vibration in hand tools or process equipment for the worker to feel it in hands, arms, or whole body?

☐ Is there so much air contaminant in the process that it settles on displays, making them difficult to see?

☐ Is the job designed so that left-handed people can do it as easily as right-handed people?

Appendix 2
Workstation Checklist

Workstation Characteristics

- ☐ Can the worker keep horizontal stretches within the range of normal arm reach? (Reach should not exceed 16–18 in.(40.6 -45.7 cm).

- ☐ Is there adequate space at the workstation to perform the work comfortably?

- ☐ Is clearance space in the workstation adequate for handling and maintenance tasks?

- ☐ Is the workstation accessible to material handling equipment?

- ☐ Does the positioning of equipment, controls, and work surface make it possible to maintain a comfortable posture?

- ☐ Is it possible for the worker to alternate sitting and standing when performing the task?

- ☐ If a chair is provided, is its design satisfactory (adequate back support, vertical adjustability, etc.)?

- ☐ Does the height of the work surface permit satisfactory arm posture? (Correct hand height is 2–6 in. (5.1 to 15.2 cm) below elbow height for most jobs.)

- ☐ If the work height is unsatisfactory, is it due to:
    - ☐ Machine
    - ☐ Work surface
    - ☐ Controls?

- ☐ Does the height of the work surface permit a comfortable view of the job being done?

- ☐ Is the height of the work surface adjustable?

- ☐ Is the texture of the work surface comfortable, taking into account hardness, elasticity, color, and smoothness?

- ☐ If pedals are used, are they positioned comfortably?

- ☐ Are pedals a comfortable size? If pedals are used, are they limited to two?

- ☐ Is the use of pedals required only on jobs performed while seated?

- ☐ Are hand controls designed to take into account the amount and types of force required to operate them?

- ☐ Are footrests and/or supports for hands, arms, and back available, if needed?

☐ If containers are used, are they placed conveniently?

☐ Are containers designed for easy maintenance and repair?

☐ Does the design of the equipment allow for easy access for maintenance and repair?

☐ Is the level of vibration low enough to avoid adverse affects on the worker?

☐ Is the workstation floor clear of clutter and obstructions that could create the risk of slips, trips, or falls?

Physical Demands

☐ Does the task require strenuous two-handed lifting?

    ☐ Lifting at too great a horizontal distance?

    ☐ Lifting more than once per minute?

    ☐ Lifting over too great a vertical distance?

☐ Does the task require strenuous one-handed lifting and reaching (such as too long a reach when feeding parts into a machine)?

☐ Are lifts awkward because they are near the floor, above the shoulders, or too far from the body?

☐ Does the job require twisting while lifting?

☐ Must the worker handle difficult-to-grasp items? (Are the items difficult to reach? Is the handhold poor?)

☐ Does the job require continual manual handling of materials?

☐ Does the job require handling of oversized objects?

☐ Does the job require two-person lifting?

☐ Must force be exerted in an awkward position (e.g., to the side, overhead, or at extended reaches)?

☐ Is help for heavy lifting or exerting force unavailable?

☐ Does the job involve peak loads of muscular effort? How often do peak loads occur? How long do they last?

☐ Can the job be designed to alternate periods of exertion and rest?

☐ Can the job be designed to alternate periods of static effort and movement?

☐ Is the pace of material handling determined by a machine (feeding machine conveyors, etc.)?

☐ Does the job lack material handling aids such as air hoists or scissors tables?

☐ Does the job involve static muscle loading (such as holding or carrying)?

☐ Is there a high level of hand tool vibration?

☐ Must the worker stand on a hard surface for 45% or more of the work shift?

☐ Is there frequent daily stair or ladder climbing?

Perceptual Load

☐ Is the illumination not satisfactory for the task?

☐ Is contrast poor between the workstation and its surroundings?

☐ Is glare present in the workstation? (If there is glare, what is its source?)

☐ Does the task require fine visual judgments? (This includes the need to detect small defects, judging distances accurately, etc.)

☐ Are controls, instruments, and equipment placed where they are difficult to see (at a bad angle, too high, too low)?

☐ Are controls, instruments, and equipment poorly lit?

☐ If warning lights are present, are they located out of the center of the field of vision?

☐ If there are auditory signals, are they difficult to distinguish from one another?

☐ Are some auditory signals hard to hear?

☐ Does the noise level prevent verbal communication?

☐ Is there a need to tell the difference between parts by touch?

☐ Is it difficult to recognize controls and tools by touch and/or position?

☐ Where dials, instruments, or displays are in use, are they difficult to read?

☐ Are dials and instruments difficult to read quickly and accurately?

☐ Is the information on the displays difficult to read from the required reading distance?

☐ Is the workstation so poorly lit that there are great differences between brightness levels in panels, dials, and surroundings?

☐ Is glare from displays a problem?

☐ Are dials grouped inconveniently?

☐ Is it difficult to differentiate among dials in a similar category because of location or lack of color coding?

☐ Are displays or dials not located near the corresponding control?

☐ Are the most important or most frequently used instruments not in the best position within the field of vision?

☐ Are the most frequently used instruments not grouped together in the same area of the field of vision?

☐ Are controls difficult to reach and operate?

☐ Are controls not standardized on similar equipment?

☐ Are there more controls than needed to perform the job?

☐ Does reading the instruments require a lot of head or body movement?

☐ Does the design of any instrument increase reading errors? Is the dial too complex for the level of information required?

☐ Are dials arranged out of the order in which they must be read?

☐ When all readings are correct, do the pointers in a group of dials point in different directions?

☐ Is it difficult to see immediately how a control is set?

☐ Does the worker's hand obstruct the dial when operating controls?

Mental Load

☐ Is the task very complex?

☐ Is the job so complex that it takes a long time to train workers?

☐ Does the task require a great deal of accuracy?

☐ Does the worker have to evaluate data before taking action?

☐ Are standards of comparison lacking?

☐ Is it difficult to recognize controls by shape, size, labeling, or color? Is this a problem in normal use? Could it be a problem in an emergency?

☐ Is the information available about performance of the job task ambiguous and irrelevant?

☐ Does the information come in faster than the worker is likely to be able to assimilate it?

☐ Must the worker keep track of several different types of information and signals at the same time?

☐ Does the job make heavy demands on short-term memory?

☐ Is the rate of information heavy enough to overload the worker?

☐ Do signals come in when the worker is concentrating on something else?

☐ Can signals from different sources occur simultaneously?

☐ Does the worker have to make a choice in response to a signal?

☐ If the worker must make a choice, does he or she know immediately if the choice was wrong? Must the job be performed within a specific time frame (e.g., when a job is paced by a machine)?

☐ Is insufficient time allowed in machine or process cycles for decisions to be made and actions to be taken?

☐ Is the job monotonous, repetitive, or unvarying?

☐ Does the job involve critical tasks with high accountability and little or no tolerance for error?

☐ Must too much information be handled in too short a time?

Work Environment

☐ Are there noise levels that interfere with conversation or performing the job?

☐ Is the noise level high enough to cause hearing loss?

☐ Is the temperature or humidity frequently uncomfortable enough to interfere with the job?

☐ Is air circulation too low?

☐ Is there too much air movement?

☐ Are workers exposed to rapid environmental changes?

☐ Are workers exposed to gaseous or vapor contaminants?

☐ Are suspended dust, mists, and other particulates present in the air?

☐ Are there wet locations that may produce shock hazards for work with electrically powered equipment?

☐ Are floors uneven?

☐ Are floors slippery?

☐ Is housekeeping poor?

☐ Is lighting inadequate for the job?

☐ Does glare interfere with reading, inspecting, and the like?

☐ Are there hot surfaces that may cause burns?

☐ Are there conditions that should require personal protective clothing or equipment?

Appendix 3
Sample Questions for Worker Evaluation of Chairs

1. Your name _____

2. Name of chair _____

3. How long do you use this chair? _____ hours.

4. Did you make any adjustments before or while using the chair?___ yes ___ no
If yes, please describe _____
_____

5. Please rate the following features of the chair by circling a number for each:

		Very comfortable	Very uncomfortable	Very good	Very bad
a.	Chair height	1	2	3	4
b.	Clearance for feet and calves under chair	1	2	3	4
c.	Seat comfort	1	2	3	4
d.	Backrest comfort	1	2	3	4
e.	Overall chair comfort	1	2	3	4
f.	Maneuverability	1	2	3	4
g.	Ease of adjustment	1	2	3	4
h.	Size of chair	1	2	3	4
i.	General appearance	1	2	3	4

6. What do you like best about this chair? _____
_____

7. What do you like least about the chair? _____
_____

8. Describe the type of work you are performing while using this chair.
Typing ___ hours   VDT use ___ hours   Clerical ____ hours   Other ___hours

Appendix 4
Checklist for Reduction of Potential Hand Tool Injuries

I. General Hand Tool (Power and Nonpower) Design Considerations
   A. Tool weight and size
      1. Keep weight of one-handed tools to less than 4 pounds.
      2. Provide places for both hands to grasp heavy tools.
      3. Heavy tools should be suspended from a tool balance device.
      4. Tools should be as compact as possible.
      5. Tools should be balanced.
      6. Grasping surfaces should be slip-resistant.
      7. Whenever possible, round the edges and corners of tools.
      8. Design tools to be used by either hand:
         a. Allows workers to alternate hands.
         b. Allows the tool to be in a worker's preferred hand.
   B. Handle thickness
      1. For a power grip, larger diameter handles are better.
         a. Optimal thickness is 1 to 1 $\frac{1}{2}$ in. (2.5 to 3.8 cm).
         b. Hand strength can be reduced up to 30% when wearing gloves.
         c. If the diameter is too large, the fingers do not overlap, there is no "locking," and strain is sharply increased. If the diameter is too small, there is an insufficient friction area and the handle cuts into the hand.
         d. T-handles should be about 1 in. (2.5 cm) thick.
      2. Length of handles
         a. The handles should be long enough so that they do not end in the palm, especially pliers.
         b. Without gloves, handles should be 4 in. (10.2 cm) minimum.
         c. With gloves, handles should be 5 in. (12.7 cm) minimum.
      3. Handle surface
         a. Nonconductive

      1) Handles should not conduct electricity or heat.

      2) Wood is the best material for two reasons:

         a) Wood releases heat to the hand more slowly than plastic or metal and so it can be held for a longer time before injury occurs.

         b) Wood gains heat more slowly than plastic ormetal, so it is less likely to reach a high temperature.

   b. Compressible—Just as a compressible floor (wood or carpet) is easier on the feet and legs than non-compressible concrete, a compressible handle material is easier on the hand.

      1) Wood is the best material

      2) Compressible rubber or plastic are also acceptable.

      3) Plastic or tape also makes the handle compressible and non-conductive.

   c. Dip metal handles in plastic or wrap with tape to cover sharp edges.

   d. Textured surfaces aid grip.

   e. Handle serrations (finger grips) cut into the fingers.

II. Hand Tool Posture Considerations

  A. Bend the tool, not the wrist

    1. Tendon movement while the wrist is not bent is less injurious.

    2. The most comfortable hand position is the "handshake" position.

    3. Another alternative to changing the tool angle is to change the orientation of the work.

  B. Use the appropriate muscle group

    1. Hand-closing muscles are stronger than hand-opening muscles.

    2. Use a spring to open hand tool blades.

Appendix 5
Ergonomics Program Development

1. Develop an action plan that is:
   - A. Easy to implement
   - B. Oriented toward results
   - C. Cost effective.
2. Set up a committee with coordinator that includes:
   - A. Plant safety personnel
   - B. Medical personnel
   - C. Plant design and industrial engineering personnel
   - D. Maintenance personnel
   - E. Industrial relations personnel
   - F. Production departments
   - G. Union/labor safety representatives
   - H. Line workers
   - I. Ergonomics expert.
3. Collect data through:
   - A. Records review
   - B. Job/task analysis
   - C. Patterns and trends of CTDs
   - D. Employee suggestions, grievances, and complaints.
4. Develop written plans and programs
5. Establish priorities for addressing problems
6. Propose solutions
7. Analyze effectiveness and feasibility of proposed solutions
8. Implement solutions
9. Monitor results and adjust as needed
10. Monitor ergonomics program
11. Strive for continuity:
    - A. Establish routine meeting dates and times.
    - B. Keep to 45–60 minutes.
    - C. Do not cancel meetings.

        D. Have alternate committee members.
12. Return on investment should include:
        A. Reduced accidents and injuries
        B. Gains in productivity
        C. Energy savings
        D. Savings from product modification.
13. Maintain cost-benefit log

Appendix 6
Selecting an Ergonomics Consultant

Ergonomics is a multidisciplinary field; therefore, your objective is to look for competency in the discipline that deals with the problem(s) you wish to solve. For example, designing control systems to minimize errors requires different skills than altering tasks to minimize cumulative trauma disorders. The background of the consultant may also affect the approach to solving the problem.

Because problems and consultant capabilities are so diverse, one table of criteria would be inadequate. Our aim is to help you evaluate the match between the problem and the candidate through suggested questions around six "Ps": problem, performance, product, process, personal, and price.

PROBLEM

Here are some problems ergonomists can address. You may be dealing with one not on this list. In any case, you should understand the problem and the desired result, and communicate these to the candidates.

- First, define the area of need—cumulative trauma disorder reduction, back injury reduction (usually involving manual materials handling), product design for comfort and ease of use, VDT (video display terminal) workstation design, and control system/panel design (for comfort and error reduction) are possibilities.

- Second, decide whether you want the consultant to solve a particular problem, sell the need for an ergonomics program to the organization (management, engineers, workers, supervisors), develop an internal program, recommend a process, recommend structure and culture changes, facilitate a program, or something else.

- If you are uncertain, perhaps you want help in defining true needs at your location. Some possible requests include:
    1. Assessment of site(s) for CTD (cumulative trauma disorder) risk factors
    2. Recommendations for corrections on jobs where CTDs are occurring

3. Response to potential or actual citation
4. Recommendations for an equitable remediation plan
5. Response to employee complaints
6. Response to union complaints
7. Development of an internal action program for avoiding CTDs and improving worker comfort/efficiency
8. Development of a program for improved quality and reduced errors
9. Training of company personnel in ergonomics
10. Training engineers to design jobs to minimize CTD risk factors and error potential
11. Testimony as an "expert witness."

PERFORMANCE

- What ergonomics problems has this person worked on?
- Who has this person worked for in the past? Are references available? Are you able to contact the right person to confirm these references?
- What level of responsibility did this person work at and what were the results? What programs resulted? Are samples available?
- Has the person been involved with regulatory agency cases? For whom? What functions? What were the results?
- Will the person be available when needed, or to meet your schedule?

PRODUCT

- Will there be a lasting product of increased competency in your organization, or only a solution to a specific problem?
- Is the focus on complex or simple solutions (e.g., automation vs. improving existing tools)?
- Is the focus on changing workstations or on improving people performance (e.g., adjustable tables vs. conditioning exercises)? Note—changes to workstations are preferred because of a higher success rate and durability of results.
- Will error reduction be included? Is this desired by you?

PROCESS

- Does it include worker input?
  —Pain/discomfort surveys
  —Problem identification
  —Problem solutions.
- Is there a training component? Will people learn to solve problems?
- Is there an organizational structure component? Will it define how to manage ergonomics?
- Is there a component to reinforce management resolve?
- Is there recognition of potential cost and quality improvements?
- What is the scope?
  —Back injury
  —Upper body CTDs
  —Whole body CTDs
  —Environmental stress (heat, noise)
  —Illumination
  —Error prevention (displays, controls, stereotypes)
  —Other.
- Will the consultant have access to all needed levels of the organization (e.g., management, supervisors, union, line workers)?
- Will needed research be defined?

PERSONAL

- What background and training does this person have (ergonomics, medical, psychology, physiology, safety, physical therapy, engineering)?
- Does this person understand how the body functions? About human behavior?
- How will background, training, education, and understanding influence the strategy for solutions?
- What is your impression of the person's ability to interface with people, management, unions, regulators? Are interpersonal skills and organizational awareness evident?
- Can this person do "hands on" field work?

- How does this person relate to you? (What are your "vibrations"?) Do you feel a sense of commitment and caring?
- Does this person appear knowledgeable about:
  —Regulations
  —Business considerations
  —"Selling" the solution
  —Training methods?
- Can this person develop rapport with your medical support people?
- Can needed medical information be accessed?
- Will this person meet any legal requirements you may have?
- How does this person stay current in the field (professional affiliations, classes, etc.)?

PRICE

- What level of commitment is there to meeting your needs? Will there be accessibility away from your location (phone, fax)? Will phone conversations be billed to you? What backups (availability, alternate skills) are available, and at what cost?
- Is the cost of services within the "usual and customary" range? If not, why? (Is there more value—or less?)
- Will the problem solutions be too expensive?

EDUCATION

If a resume or curriculum vita is submitted, here are some checkpoints:

- What type of degree? Preferred areas of concentration would be ergonomics, specifically, or related fields such as industrial engineering or occupational health.
- What specific courses and how many?
- What types of continuing professional education—seminars, correspondence courses, and so on?

EXPERIENCE

- Positions held (with description of jobs)?
- Percentage of time involved in ergonomics?
- Loss experience of the companies for which the person worked?
- Specific types of ergonomic problems dealt with?
- Types of action taken to address these problems?
- Success of these actions?
- Description of major components of a comprehensive ergonomics program?
- Experience in developing and presenting ergonomics training programs?

Selected Glossary

Ergonomics. Ergonomics is the study of the design of requirements of work in relation to the physical and psychological capabilities and limitations of people; that is, ergonomics seeks to fit the job to the person rather than the person to the job.

Ergonomic hazards. Ergonomic hazards refer to a combination of stressors or workplace conditions that may cause harm to the worker. Improperly designed workstations, tools, and equipment; improper work methods; and excessive tool or equipment vibration are examples of this type of hazard. Other examples stem from job and process design problems that include aspects of work flow, line speed, posture, force required, work/rest regimens, and repetition rates.

Ergonomic disorders. Ergonomic disorders (EDs) are the range of health disorders arising from repeated stress to the body due to exposures to ergonomic hazards. These disorders may affect the musculoskeletal, nervous, and neurovascular systems. EDs include the various occupationally induced cumulative trauma disorders, cumulative stress injuries, and repetitive motion disorders.

A main distinction between EDs and strain or sprain injuries is that the latter usually result from a single act, such as acute trauma. EDs, on the other hand, develop gradually over periods of weeks, months, and years, and there are few, if any, distinctive or dramatic features surrounding their onset. EDs include damage to the tendons, tendon sheaths, synovial lubrication of the tendon sheaths, bones, muscles, and nerves of the hands, wrists, elbows, shoulders, necks, backs, and legs. Some of the more frequently occurring occupationally induced EDs include chronic back pain, carpal tunnel syndrome, DeQuervain's disease, epicondylitis (tennis elbow), Raynaud's syndrome (white finger), synovitis, stenosing tenosynovitis crepitans (trigger finger), tendonitis, and tenosynovitis.

Ergonomic stressors. Ergonomic stressors are regarded as synergistic elements or functional subunits of ergonomic hazards, of which one or more can combine to constitute an ergonomic hazard. Exposure to jobs, operations, processes, or workstations that have multiple stressors decreases the latency of the onset of ergonomic disorders, depending upon the relative intensity, duration, frequency, and combination of each stressor contributing to the ergonomic hazard. Examples of ergonomic stressors include repetitiveness of activity or motion; excessive or required force or grip in performing an activity; awkward, static, or prolonged positioning of the body; vibration; and lighting conditions.

Systematic approach or systems approach. A systems approach to safety and health management means a comprehensive program by the employer that addresses workplace processes, operations, and conditions as interdependent systems to identify, evaluate, prevent, eliminate, or reduce all types of hazards to employees. A comprehensive program to address complex problems, such as the presence of a variety of ergonomic stressors in the workplace, may require the integration of a combination of solutions.

Systematic analysis. Systematic analysis is a holistic, quantitative, and qualitative evaluation process for investigating ergonomic stressors and hazards in the workplace. This process describes and evaluates employee exposures to equipment, jobs, processes, workstations, and tasks, and determines the extent to which exposures are or may become hazardous. In ergonomics, this analysis focuses on the stressors that result from the relationship between employees, their jobs, and the work processes. Systematic analysis also includes determining a plan for controlling those hazards.

A systematic analysis of exposures to ergonomic stressors is initiated by identifying tasks that, based on an assessment of company records, employee surveys, and/or surveillance, require further hazard characterization. Specific behaviors required of persons performing a task, job, process, or operation are then observed, described, and evaluated. This analysis includes the examination of the relationship between the job in question and other job-units in the production process, so that, for example, alternate methods of production may be substituted for the hazardous one. The objective of this step-by-step analysis is to determine if and where the limits of human capability have been exceeded. Techniques of step-by-step analysis include those commonly used in industrial engineering and system safety engineering, such as fault tree analysis or job hazard analysis.

Ergonomic surveillance. Ergonomic surveillance is the ongoing systematic collection, assessment, and interpretation of health, incidence, and exposure data in the process of describing and monitoring the circumstances which may be related to ergonomic hazards or the presence thereof. Ergonomic surveillance data, augmented by other sources of information, can be used to assess the need for occupational safety and health action and to assist in planning, implementing, and evaluating ergonomic programs and early interventions.

Health management. Health management, often referred to as medical management, is a component of a managerial system approach to ensure early identification, evaluation, and treatment of signs and symptoms of ergonomic disorders and to aid in the prevention of ergonomic disorders or symptoms. Health management is broader in scope than medial surveillance provisions in other OSHA standards. While health management encompasses traditional medical surveillance, it also involves, among other things, the participation of trained health care providers in periodic workplace walk-throughs and follow-up assessments of workers who have developed signs and symptoms of ergonomic disorders. (From *Federal Register*, vol. 57, no. 149, August 3, 1992, U.S. Department of Labor, Occupational Safety and Health Administration.)

Bibliography

Armstrong TJ. *Ergonomics Guides: An Ergonomics Guide to Carpal Tunnel Syndrome*. Akron OH: American Industrial Hygiene Association, 1983.

Chaffin DB and Andersson GB. *Occupation Biomechanics*. 2nd ed. New York: John Wiley and Sons, 1991.

Cook TM, Mann S, Lovested GE. Dynamic comparison of the two-hand stoop and assisted one-hand lift methods. *Journal of Safety Research*. 21:53–59, 1980.

Eastman Kodak Company, Ergonomics Group. *Ergonomic Design for People at Work*, vols 1 & 2. New York: Van Nostrand Reinhold, 1989.

Grandjean E. *Fitting the Task to the Man*. 4th ed. London: Taylor and Francis, Ltd, 1988.

Habes DJ and Putz-Anderson V. *The NIOSH program for evaluating biomechanical hazards in the workplace.* Journal of Safety Research. 16:49–60, 1985.

National Institute for Occupational Safety and Health. *Work Practices Guide for Manual Lifting*. DHHS (NIOSH) Publication No. 81–122. Washington, DC: Government Printing Office, 1981.

National Safety Council. *Accident Prevention Manual for Business & Industry*, vols 1 & 2. 10th ed. Chicago: National Safety Council, 1992.

———. *Fundamentals of Industrial Hygiene*. 3rd ed. Chicago: National Safety Council, 1987.

National Safety Council (Chicago)/Swedish Work Environment Fund (Stockholm, Sweden). *Making the Job Easier . . . An Ergonomics Idea Book*. Chicago: National Safety Council, 1988.

Putz-Anderson F. *Cumulative Trauma Disorders: A Manual for Musculoskeletal Diseases of the Upper Limbs.* Philadelphia: Taylor & Francis, 1988.

Snook S, Ciriello V. The design of manual handling tasks: revised tables of maximum acceptable weights and forces. *Ergonomics.* 34(9):1991.

U.S. Department of Labor, Occupational Safety and Health Administration. *Ergonomics Program Management Guidelines for Meatpacking Plants.* OSHA 3123, 1990.

Van Cott HP and Kinkade RG, eds. *Human Engineering Guide to Equipment Design.* Washington, DC: U.S. Government Printing Office, 1972.

Please contact the National Safety Council, 1121 Spring Lake Road, Itasca, IL 60143-3201, for its publications and audiovisual materials on ergonomics and related topics.

Index

A
Absenteeism, in identifying ergonomic problems, 11, 12
Accidents, analysis of, in identifying ergonomic problems, 12
Action Limit (AL), 49
Acute fatigue, 31
Anthropometry
    data in, 27–30
    definition of, 6, 27
    evaluating and designing workstations, 30
Assisted one-hand lift method, 57–59
Atmospheric pressure stress, 33

B
Back, structure of, 41–42
Back injuries, causes of, 41–43
Back stress, in seated work, 37–38, 43
Board of Certification in Professional Ergonomics (BCPE), 2, 7–8
Bursitis, 64

C
Carpal tunnel syndrome (CTS), 67–75
    causes of, 67
    preventive measures for, 67
        force, 69–70
        grip, 69
        job rotation, 74
        mechanical stresses, 71–73
        posture, 67–69
        repetitive motion, 69
        stress, 69
        temperature, 69
        vibration, 69
    symptoms of, 67
Carrying tasks, redesigning, 55–56
Certification, in professional ergonomics, 7–8
Certified Human Factors Professionals (CHFP), 8
Certified Professional Ergonomists (CPE), 8
Chairs
    design of, in seated work, 37–38
    sample questions for worker evaluation of, 101
Chemical stressors, 33
Chronic fatigue, 31

C
Circadian rhythm, disruptions in, 35–36
Controls, guidelines for, 82–83
Cumulative fatigue, 31–32
Cumulative trauma disorders (CTD)
    bursitis, 64
    carpal tunnel syndrome (CTS), 67–75
        causes of, 67
        preventive measures for, 67
        symptoms of, 67
    common risk factors, 61–63
    De Quervain's stenosing tendinitis, 64
    epicondylitis, 65–66
    ganglion cysts, 64
    hand-arm vibration syndrome (HAVS), 65
    hypothenar hammer syndrome, 65
    incidence of, in identifying ergonomic problems, 11
    rotator cuff syndrome, 64
    tendinitis, 64
    tenosynovitis, 64
    thoracic outlet syndrome, 64
    trigger-finger syndrome, 65

D
Danger, as psychological stressor, 34–35
Data, anthropometric, 27–30
De Quervain's stenosing tendinitis, 64
Disabled person, ergonomic adjustments for, 16
Displays, guidelines for, 83–84
Drawing, in ergonomic task analysis, 24
Driving posture, 43

E
Employee complaints, in identifying ergonomic problems, 11, 13
Employee-generated changes in the workplace, in identifying ergonomic problems, 11, 13
Employee, involvement of, in implementation of ergonomics program, 86–87
Environmental factors, in lifting injuries, 46–47

119